INNOVATION REBOOT

INNOVATION REBOOT

How to build, manage and assess innovation capability in organisations and teams

CHRIS HAKES

Published by:

Leadership agenda

Innovation Reboot™ is a registered trademark of the non-profit Innovation Reboot project launched by Leadership Agenda Limited in 2012.

Copyright © 2012 by Chris Hakes. All rights reserved.

Published by Leadership Agenda Limited on behalf of the Innovation Reboot Project,

Somersham, IP8 4QA, UK. www.leadershipagenda.org and www.innovationreboot.org

First printed edition published 1st July 2013.

The right of Chris Hakes to be identified as author of this Work has been asserted by him in accordance with the UK Copyright, Designs and Patents Act 1988.

Innovation Reboot is a registered trademark, UK Patent Office reference: UK00002648212 and the term should be acknowledged as such, in the format of Innovation Reboot™ when used outside of this book.

Apart from any fair dealing for the purposes of research or private study, or criticism or review, as permitted under the UK Copyright Design and Patents Act, 1988, and the sharing under Creative Commons, of the Innovation Capability Framework diagram (provided it is cited, in all cases of use, as described in the 'We Are What We Share' section of this book), no part of this publication may be reproduced, stored in a retrieval system, or transmitted by any means – electronic, mechanical, photographic (photocopying), recording or otherwise – without prior permission, in writing, from the publisher or payment of per copy fees to the Copyright Licensing Agency in the UK, the Copyright Clearance Centre in the USA, or in accordance with the terms of licences issued by the appropriate Reproduction Rights Organisation outside of these territories.

The publisher and author have used their best efforts to make sure the information in this book is accurate and up to date, but it is a general guide only. The publisher and author specifically do not warrant any of the techniques described as being fit for a particular purpose. Before acting you should seek advice from a relevant professional advisor who can consider your individual circumstances. Websites that are offered as sources for further information in this book may have changed between the time of writing and of reading this book and cannot be relied on for other than general advice, when available. The publisher and author take no responsibility for any loss, damage or injury resulting from use of techniques described in this book.

British Library Cataloguing in Publication Data:
A catalogue record for this book is available from the British Library.

ISBN: 978-1-904861-02-7 (cloth)

CONTENTS

Preface and Ideas on How to Use this Book   11

Section 1: Basics and Ambitions   17

Chapter 1
Definition and Scope of Innovation   19

Chapter 2
Common Imperatives   25

Chapter 3
Revenue and Capability Gaps   31

Chapter 4
Capability Development and Innovation Governance   35

Section 2: Capability Assessment and Management — 41

Chapter 5
Element 1 Innovation Strategy — 45
   Element 1.1: Integration to Core Business Strategy — 47
   Element 1.2: Focussed Ambitions — 48
   Element 1.3: Assumptions, Hypotheses and Triggers — 57
   Capability Assessment – Innovation Strategy — 59

Chapter 6
Element 2 Data and Analytics — 65
   Element 2.1: Broad Horizons – Inputs — 67
   Element 2.2: Analysing, Connecting and Forecasting — 69
   Element 2.3: Avoiding Decision-Making Bias — 72
   Element 2.4: Systematically Share Emerging Analyses — 73
   Capability Assessment – Data and Analytics — 77

Chapter 7
Element 3 Ideate – Create or Acquire Ideas — 83
   Element 3.1: Creativity from Inside the Organisation — 85
   Element 3.2: Obtaining Ideas from Outside — 87
   Element 3.3: Ideas Management — 88
   Element 3.4: Selecting Likely 'Winners' to further validate — 89
   Capability Assessment – Ideation — 95

Chapter 8
Element 4 Validate Key Ideas and Make Business Plans — 101
   Element 4.1: Phased, Rapid, Validation — 103
   Element 4.2: Efficient and Timely Business Planning — 105
   Capability Assessment – Validation — 111

Chapter 9
Element 5 Scale the Best New Ideas or Businesses — 117
   Element 5.1: Processes – How much glue? — 119
   Element 5.2: Getting the Right People — 121
   Element 5.3: Hyper-Agility at Launch — 125
   Capability Assessment – Scaling — 127

Chapter 10
Element 6 Innovation Ready Culture — 133
    Element 6.1: Making the Innovation Imperatives Clear — 135
    Element 6.2: Developing Competencies and Opportunities — 136
    Element 6.3: Providing Personal Feedback to Everyone — 137
    Capability Assessment – Innovation Ready Culture — 139

Chapter 11
Element 7 Develop Future Focussed Leaders — 145
    Element 7.1: Leadership Ambidexterity — 147
    Element 7.2: A Collaborative Style, When Needed — 148
    Element 7.3: Manage Innovation Enabling Activities — 150
    Capability Assessment – Future Focussed Leaders — 155

Appendices — 161

We Are What We Share — 163
What Next? — 165
Notes — 167
Glossary of Terms — 171
Acknowledgements — 177
About the Author — 179
Index — 181

List of Figures

Figure 0.1 Innovation Capability Assessment Framework　14

Figure 1.1 Innovation 'sparks' at the Intersection　22
Figure 1.2 A Simple Process View of Innovation　23

Figure 2.1 2060 Global GDP – Predicted Share by Country　26
Figure 2.2 Speed 'Records' for Scaling and Diffusion　29

Figure 3.1 Using a Revenue Forecast to Identify Innovation Gaps　32

Figure 4.1 Innovation Capability Assessment Framework Overview　36
Figure 4.2 Logic of the Innovation Capability Assessment Framework　37

Figure 5.1 Innovation Capability Framework: Innovation Strategy　46
Figure 5.2 Opportunity Analysis Matrix – Two dimensions　49
Figure 5.3 Opportunity Analysis Matrix –16 Potential Investments　50
Figure 5.4 Opportunity Analysis Matrix – Strategic categories　51
Figure 5.5 Opportunity Analysis Matrix – Level of Risk/Opportunity　53

Figure 6.1 Innovation Capability Framework: Data and Analytics　66
Figure 6.2 Segmentation of Insight Opportunities　68
Figure 6.3 The 'Road' from Data to Foresight is Long　70
Figure 6.4 Create Opportunity Briefs in Areas of Strategic Interest　74

Figure 7.1 Innovation Capability Framework: Ideation　84
Figure 7.2 Wise Innovators use a Range of Techniques　86
Figure 7.3 Wise Innovators Keep Ideas Registers　88
Figure 7.4 Create Validation Plans　90
Figure 7.5 Wise Innovators know their Idea Conversion Ratios　91
Figure 7.6 Idea Evaluations in a Simple, Ease – Impact Matrix　92

Figure 8.1 Innovation Capability Framework: Validate and Plan　102
Figure 8.2 Validation Occurs in Phases　103
Figure 8.3 Create Simple but Sound Business Plans　106
Figure 8.4 A Sound Business Plan Should Pass Three Tests　107
Figure 8.5 Innovation Diffusion and Take-Up over Time　109

Figure 9.1 Innovation Capability Framework: Scaling　118
Figure 9.2 Critical Mass Trigger Points in Organisational Growth　120
Figure 9.3 Managing Team Expectations　123

Figure 10.1 Innovation Capability Framework: Innovation Ready Culture　134
Figure 10.2 Building a Supportive Culture　135

Figure 11.1 Innovation Capability Framework: Future Focussed Leaders　146
Figure 11.2 Transitions in a Future Focussed Leader　147
Figure 11.3 Levels of Participation and Collaboration　148

To my early mentor, Ted Poynter, for gracefully pointing out my many defects and supporting me to overcome some of them.

PREFACE and IDEAS ON HOW TO USE THIS BOOK

First, thank you very much for purchasing this book. Half of the author's royalties from the sales of this book will go to support the non-profit Innovation Reboot™ project at www.innovationreboot.org, of which, more later. Your support allows this experiment to proceed.

This book was born out of personal experiences that led me to observe that, all too often, although Innovation is recognised as important in the 'C-Suite' or Boardroom, there is typically a massive gap between Innovation rhetoric and action. Most senior managers recognise the role that Innovation can play in developing their organisations; they know that Innovation is often essential to creating or maintaining revenues and value, but they can become paralysed by past failures or a lack of consensus on how best to manage it'.

My personal experience of this dilemma started in my early working life, as a businessman and entrepreneur, and has been reinforced through a long association with the EFQM Excellence Award (www.efqm.org) and other similar global, national and regional business award schemes, where I have had the fantastic privilege of working with and assessing the performance of many excellent, award-winning organisations. When supporting or assessing potential award winners (for EFQM and others) I always set myself a 'mission' to ensure that at least some of the time was spent on debating Innovation enabling practice and challenges. The insights obtained from these interactions have become a significant part of this book and I've widely quoted the views of so-called 'wise innovators' in subsequent chapters.

In searching wider for Innovation good practice, research and examples to share, I soon came to the conclusion that academia was not, currently, going to provide many answers to boardroom inaction on Innovation. Most academic work on Innovation does little to help; valid research is fragmented and where systemic views exist they are rooted in the micro-perspective of a specific discipline, and little cross-faculty, cross-discipline work is done to consider the broader subject from the perspective of holistic organisation-wide management systems, within which most business managers have to operate. Academia, so far, perhaps constrained by how it is incentivised and funded, has not risen to the challenge of proposing practical operational guidance on what may drive effective Innovation in organisations.

This book and the related sharing site at www.innovationreboot.org are an experiment, aimed at both academia and organisational managers, with the objective of collating and maintaining a dynamic body of knowledge on what best enables Innovation in differing organisations. The logic is that readers will hopefully benefit from using the Innovation Capability Framework and insights in this book, which may lead them to wish to learn and share more about the elements of the framework (hence the website), which may in turn, then act as a catalyst to frame and encourage deeper academic research, debate and sharing networks. If I dream a little, I could even hope we may get some sponsors who may help provide resources to fund academic programmes to do this in depth.

Back to the book… You can read and use it in many ways; it is intended to be a 'smorgasbord'. You can pick and choose from the 'buffet' at any time – the chapters stand alone – but if you want the entrées before the dessert, then good reader logic is as follows:

Start with **Section 1: Basics and Ambitions**, in order to review:

> **Chapter 1: Definition and Scope of Innovation:** Are you clear on how you define 'Innovation' in your organisation and how broad the scope of the related activity should be?
>
> **Chapter 2: Common Imperatives:** Do you understand the global forces that lead most organisations to conclude that Innovation is an imperative?
>
> **Chapter 3: Revenue and Capability Gaps:** Are you clear on your strategic shortfalls, specifically any competency and revenue gaps that you may have?
>
> **Chapter 4: Capability Development and Innovation Governance:** This chapter helps you understand how to use the assessment tools in this book to create an Innovation capability development plan.

Then proceed to **Section 2: Innovation Capability Assessment and Management**, where you will be guided on how to assess your organisation's performance for each of the seven Innovation Capability Framework elements that are shown in outline in Figure 0.1.

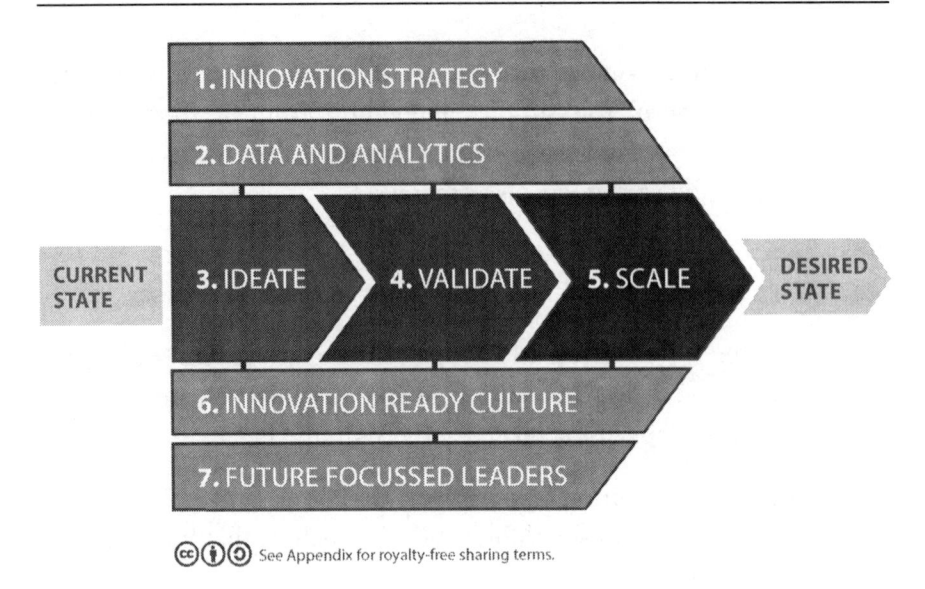

Figure 0.1 Innovation Capability Assessment Framework

```
                1. INNOVATION STRATEGY
                2. DATA AND ANALYTICS
CURRENT                                              DESIRED
STATE      3. IDEATE   4. VALIDATE   5. SCALE        STATE
                6. INNOVATION READY CULTURE
                7. FUTURE FOCUSSED LEADERS
```

See Appendix for royalty-free sharing terms.

The seven elements of this framework are described in Chapters 5 to 11, where you can read insights into what 'wise innovators' do and then follow a series of questions to help you assess your own organisation's successes with, and capabilities for, Innovation. The book is designed to help you to form your own views, record your own notes and make your personal input into an Innovation Capability Assessment of your organisation or team.

One point that I ought to raise about now is a 'health warning'. This book is no more than a collection of observations on what has worked in those who I perceived were/are wise and successful innovators. I cannot guarantee anything in this book will work for you, I do not make or share any academic research-based claims for this work, nor do I suggest exclusivity of thinking for the ideas tabled here. This book is simply a personal analysis, attempting to connect, and bring together, a summary of the practices that I've seen drive Innovation success in organisations and which may provide a framework for ongoing debate and sharing. Within the book itself I have included the minimum number of references to additional materials/reading. I have not followed the current penchant for extensive lists of pseudo-academic citations in what is intended to be, primarily, a practical business workbook for managers. Those with a hunger for more learning, references and links should go to the Innovation Reboot website, where you will find a building collection of such additional materials.

Finally, Innovation can be said to be the only possible insurance against irrelevance. I hope this book and its assessment tools may help you and/or your organisation, public or private, large or small, to insure yourself against the future. If you disagree with anything and/or if you have great ideas to share, I hope you can find the time to tell us and contribute to the Innovation Reboot Project at www.innovationreboot.org.

Chris Hakes, Cambridge UK, July 2013

SECTION 1
BASICS and AMBITIONS

CHAPTER 1

DEFINITION and SCOPE of INNOVATION

The world seems to be full of business leaders and politicians promising, 'we will innovate our way out of….' (any challenge in which they find themselves). It's a non-contentious subject; it's easy to talk about Innovation; everyone thinks that Innovation is a good thing, no one ever challenges the aim, but many organisations subsequently do nothing. They are stalled in a gap between generalised Innovation rhetoric and action. When looking to overcome such issues an essential first step is to take a look at the definition, scope and meaning of the word 'Innovation' as used in your organisation.

DEFINING INNOVATION

Wise innovators know that Innovation success is rarely obtained by accident or good luck. Those who succeed are very clear about what Innovation is and why it is important to them. They articulate and communicate a detailed, organisationally specific, definition of Innovation so that all involved are clear on their aims and can engage with them.

Wise innovators recognise the potential breadth of Innovation. A.G. Lafley (retired CEO of Proctor and Gamble and often cited for playing a pivotal part in their past decades growth), offers an interesting insight from his time at Proctor Gamble*[1]:

'A lot of companies, have defined Innovation rather narrowly, as technology, or as a mix of product and technology. That's just the beginning for us. We try to define it, in the terms that our consumers view it and experience it. So for us Innovation is the brand in addition to the product. It's the design of the shopping and the usage experience in addition to the functional attributes or benefits. It is the business model. It is the way we go to market and the supply chain. It is the way we create a cost structure so we can deliver delightful new products at affordable costs.'

*[1] Lafley, A. G., and Martin, R. L., *'Playing to Win: How Strategy Really Works'*. First Edition. Harvard Business Review Press, 2013 and in associated HBR blogs.

Wise innovators create a definition of Innovation that is clear, easily communicable, appropriately broad and capable of being aligned to their, organisation specific, strategic aims for the use of Innovation.

But how do you do this if you have not done so before? If you are visiting Innovation for the first time, an Internet search on 'definition of Innovation' will bring many options to review; a few good examples to consider are listed below:

Dictionaries:

Innovation comes from the Latin word: Innovare – *'to make something new'*.

The Cambridge English Dictionary defines Innovation as: *'(the use of) a new idea or method'*.

The views of some global influencers/networks:

The Organisation for Economic Cooperation and Development (OECD) use the definition *'An Innovation is the implementation of a new or significantly improved product (goods or service), or process, a new marketing method, or a new organisational method in business practices, workplace organisation or external relations'*.

The European business network, EFQM (see www.efqm.org), defines Innovation as, *'The practical translation of new ideas into products, services, processes, systems or social interactions'*.

A respected 'thinker' from the past:

The Austrian economist, Joseph Schumpeter, defined Innovation, over a generation ago, with concepts that many perceive as still, in-part, valid today, as: *'The introduction of goods (product/service offerings), that are new to consumers, or of significantly higher quality/value than previously available and/or methods of production, including platforms of operation, which are new to a particular branch of industry'*.

The EFQM, OECD and Schumpeter definitions are often perceived as being particularly useful examples, to review as a basis for the development of a company's own definition, as they bring with them a statement of the scope and potential outcomes for Innovation. In the EFQM example their view is that the scope of outcomes can embrace *'products, services, processes, systems or social interactions'*.

Defining the scope of Innovation, and linking this to outcome categories, can be an important part of creating a robust company specific definition of the term.

THE SCOPE OF INNOVATION

To illustrate the potential breadth of the subject, one useful, but simplistic, way of looking at Innovation, is that it can be said to occur when an organisation matches a need – for example, a customer's need for a product feature not previously provided (think of GPS when it was first added to domestic cameras) – and then addresses it by application of capabilities that it has either created, previously built, or can readily obtain.

Innovation occurs at the intersection of 'what is needed' (some customers have a new unmet need for GPS to location fix their photographs) and 'what is possible' (the organisation has created, built or is willing to acquire GPS technology capabilities). Those who get there first can claim to have innovated.

Figure 1.1 Innovation 'sparks' at the Intersection

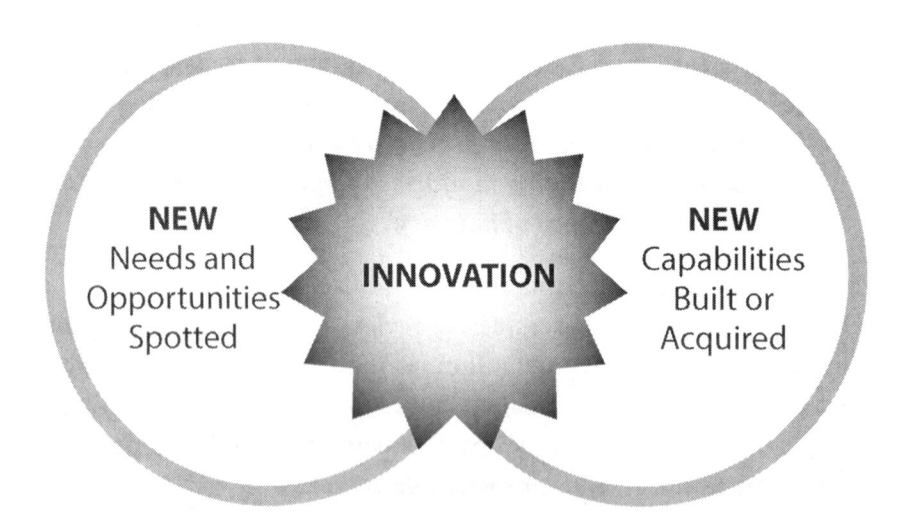

NEW Needs and Opportunities Spotted → **INNOVATION** ← **NEW** Capabilities Built or Acquired

The potential scope for Innovation is therefore as broad as the 'lens' by which the organisation seeks to spot new opportunities/needs or radically alter how it delivers on current opportunities/needs. This means that the scope of Innovation is potentially limitless and is only restrained by the capabilities an organisation has, or is willing to develop. Thinking deeper about innovation 'outcome categories' can help further define and focus Innovation.

CATEGORIES OF INNOVATION

Articulating Innovation outcome categories can be another useful way of enhancing an organisational-specific definition of Innovation, to give clarity on the potential breadth of desired outcomes and thereby indicate the potential scope of Innovation. The scope of such thinking could include a range of outcome categories, such as business model, operational or alliance processes, products, services, channels and brands.

Innovation can be seen as a simple flow process (Figure 1.2), although the reality is much more complex than this and is expanded on in Chapters 5–11 of this book, but for now, the concept of some simple categorisation of Innovation outcomes (right-hand side of Figure 1.2) is worthy of review.

Figure 1.2 A Simple Process View of Innovation

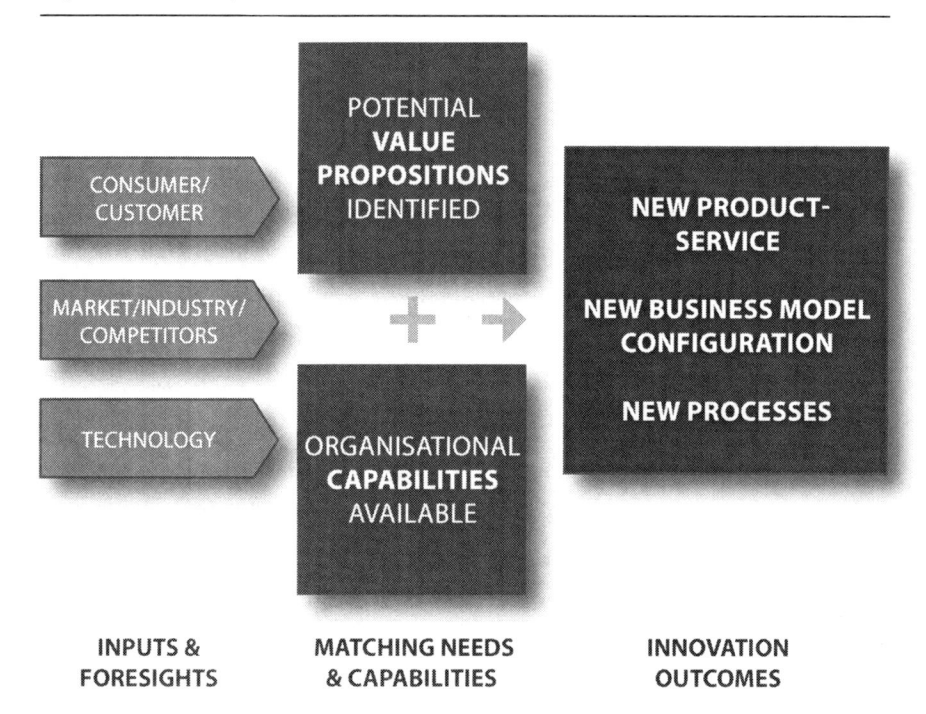

CONSUMER/ CUSTOMER

MARKET/INDUSTRY/ COMPETITORS

TECHNOLOGY

POTENTIAL VALUE PROPOSITIONS IDENTIFIED

ORGANISATIONAL CAPABILITIES AVAILABLE

NEW PRODUCT-SERVICE

NEW BUSINESS MODEL CONFIGURATION

NEW PROCESSES

INPUTS & FORESIGHTS

MATCHING NEEDS & CAPABILITIES

INNOVATION OUTCOMES

These simple categorisations of Innovation outcomes are not mutually exclusive. For example, new products are often in part delivered by new processes and can involve new business models and value capture methods. Such outcomes will also have differing natures according to the degree of Innovation achieved. Some Innovation outcomes maybe smaller incremental or 'continual' improvements, derived from new practices or knowledge. Others maybe more radical transformations, for example to a business model or brand. A wise innovator seeks a mix of all such outcomes and will ensure that the ways in which they define Innovation makes this clear.

An analysis of Amazon's dramatic and successful growth shows a mix of all such outcomes. They have moved their brand and operations into differing business sectors with differing products, processes and reconfigured business models. Analysis of their growth shows evolution from a start in 'book sales', migrating to 'publishing' (CreateSpace), expanding to 'home shopping' (with innovative low-cost home delivery) to 'electronic devices' (Kindle) and, more recently the addition of 'cloud data services' (Amazon servers support data users in around 200 countries). These transitions have been bold and embraced all three categories of Innovation outcome exampled in Figure 1.2.

CHAPTER SUMMARY

Wise innovators do not limit their Innovation scope or category thinking, they ensure that their definition of Innovation makes clear that they are (most likely) willing to embrace all categories, such as those listed on the right-hand side in Figure 1.2. Although, at any given time, their Innovation Strategy (See Chapter 5) will make clear where their current focus is and how budgets will be prioritised and spent.

Before moving to the next chapter, where you will review what makes innovation an imperative for all organisations, now may be a good time to ask yourself if you have a valid definition of Innovation, which is not limiting in its scope, that is clear, easily communicable and that is aligned to the aims you have for Innovation in your organisation/team.

CHAPTER 2
COMMON IMPERATIVES

Wise innovators know, that in order to optimise Innovation success, obtaining a shared understanding of an Innovation imperative, an understanding of why Innovation is unavoidable and essential to them, shared by all those whom they wish to engage with in the pursuit of it, is essential.

This chapter explores two unavoidable, global, strategic changes/challenges that are widely recognised as creating Innovation imperatives for any organisation, large or small, public or private, anywhere. Additionally, all wise innovators will identify their own organisational-specific revenue generation and profit imperatives for Innovation, and these are expanded on in the following chapter (Chapter 3).

COMMON IMPERATIVE #1: ONGOING GLOBALISATION

Wise innovators recognise 'standing still' is not an option because of the ever more rapidly changing ways in which economies, ecology, capital, markets, knowledge, resources, operational sites and people are linked, interdependent and accessible on a global scale. They recognise that these often accelerating changes will create both opportunities and risks that will challenge what they think they exist for today.

The likely magnitude of global change can be seen in the well respected forecasts from the 'Organisation for Economic Cooperation and Development' (OECD). Shown in Figure 2.1 are the latest OECD estimates for the foreseen country share of world Gross Domestic Product (GDP) in 2060[*1].

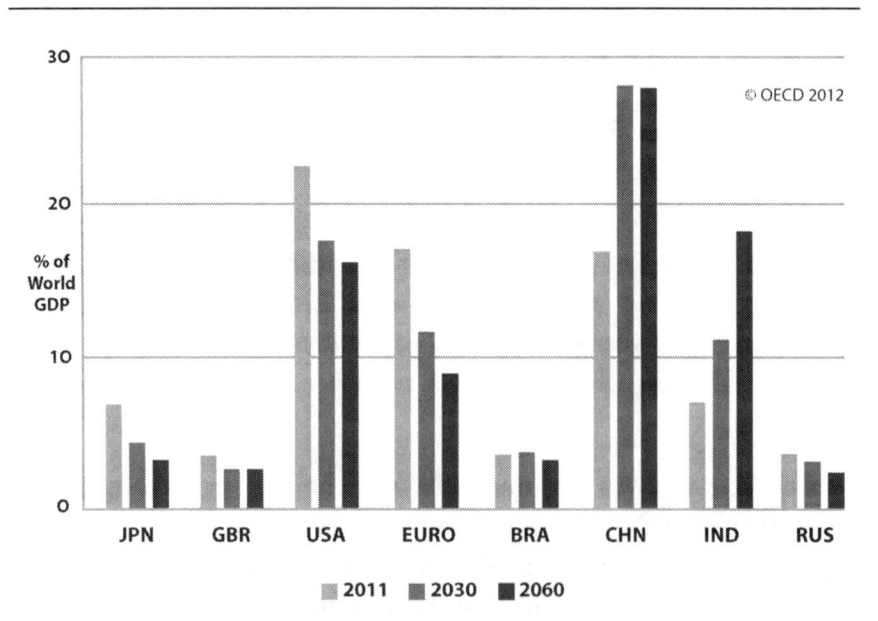

Figure 2.1 2060 Global GDP –Predicted Share by Country

JPN: Japan, GBR: Gt. Britain, USA: United States of America, EURO: Euro Zone, BRA: Brazil, CHN: China, IND: India, RUS: Russia

[*1] Looking to 2060: Long-term growth prospects for the world. *'Economic Policy Paper No. 3'*, November. 2012, OECD, Paris.

Undoubtedly, like any forecast of this nature, they are out of date the moment they are published and the reality may be significantly different, but Figure 2.1, perhaps usefully, if not precisely, summarises the foreseen general magnitude of such changes. Related OECD predictions include four bold (maybe also useful but not precise) forecasts:

1. The United States conceding its place as the world's largest economy to China, as early as 2016

2. The combined economies of India and China soon surpassing the collective economy of all the G7 nations

3. China seeing more than a sevenfold increase in per capita income over the coming half century

4. The Eurozone gradually losing ground on the global GDP table to countries with a younger population, like Indonesia and Brazil.

The world's business market place now spans geographies, economic blocks, social groups and economies in ways that it has never done before. The nature of competition is ever changing; new products and new entrants are redefining established industries; for many organisations Innovation may be the only enduring insurance against commoditisation and global price pressures.

Successful innovators know the need to act and innovate with forecasts such as above in mind. They realise that, either as opportunities (enlarged global 'playing fields') or risks (easier access for global competition), change of this order of magnitude cannot be ignored.

COMMON IMPERATIVE #2:
UBIQUITOUS TECHNOLOGY and DATA

Whilst every wise innovator will monitor the highly specific technologies that maybe core to their business, another common imperative, recognised as being universally applicable to most innovators (irrespective of the nature of their business), is the ubiquitous growth of easily accessible information/communications technology and the readily available, large volumes of data that the world can now offer.

At the time of writing IBM estimate[*2] that, every day, the world obtains an extra 2.5 quintillion bytes of data. Most of us cannot comprehend that figure (it has 17 zero's and is equivalent to 3.3 billion CD disks/day) and some will challenge it. For sure, *not* all of the data they quantify in this way is useable – even less, relevant – but the key point here is that wise innovators know that, given that such rapid changes to data availability are occurring, then data use must be part of their Innovation thinking.

[*2] If you have a mind that can handle this sort of stuff, take a look at: Zikopoulos P. C. and Eaton C., 2012, 'Understanding Big Data', IBM, McGrawHill.

Similarly, the pervasive dissemination of technology, particularly information and communication technologies, is transforming the speed and way in which people interact, engage, communicate and provide feedback. Millions of customers can now 'socially' share their views with little control or oversight from organisations that may have historically thought they could/should control or influence them.

The combination of the use of such data and technology, along with global distribution and logistic platforms, means that successful innovators can find new insights and scale new ideas to large volumes in very short time frames. Technology is allowing speed to become an increasingly important source of competitive advantage or risk and is something that wise innovators contemplate and incorporate in their Innovation strategies. It is through such speed that often a first mover will achieve the biggest gain and can lockout, at least in the short term, those who try to follow.

Figure 2.2 illustrates the ever-decreasing speed 'records' for scalability, showing how, in some sectors, scaling and diffusion times have contracted allowing some new businesses/products to reach global scale within a few days or months.

Figure 2.2 Speed 'Records' for Scaling and Diffusion[*3]

50 million users:

RADIO	38 years
INTERNET	4 years
FACEBOOK	1 year
TWITTER	9 months
GOOGLE+	3 months
ANGRY BIRDS "SPACE"	35 days

[*3] Based on updates on original data provided in the United Nations Millennium report, 'The role of the United Nations in the 21st century', 2000, United Nations, New York.

Wise innovators know the increasingly ubiquitous spread of easily accessible technology and data creates a source of Innovation opportunity or risk, which cannot be ignored. To help spot opportunities or risks, wise innovators will have a sound approach to the use of data and analytics, Chapter 6 expands on this in detail, most will have a disruptive technology 'watch-list'. The global knowledge and consultancy group, McKinsey & Company, in a recent paper[*4], share a view that there are twelve potentially disruptive technology advances that all should 'watch'. An Internet search of any of these terms will soon provide insights into the rate of change that maybe about to hit you. The McKinsey 'watch-list' (in no specific order) includes:

- Mobile Internet
- Automation of knowledge work
- Internet of things
- Cloud technology
- Advanced robotics
- Autonomous vehicles

- Next generation genomics
- Energy storage
- Additive layer manufacturing (some would say 3D printing)
- Advanced materials
- Advanced oil and gas recovery
- Renewable energy.

[4] Manyika J., Chui M., Bughin J., Dobbs R., Bisson P., and Marrs A., *'Disruptive Technologies: Advances that will transform life, business and the global economy'* McKinsey Global Institute, McKinsey Quarterly, May 2013, Stamford.

CHAPTER SUMMARY

Effective innovators are clear on the imperatives that are making Innovation essential to them. They openly share these so as to be able to obtain the support of those who they need to contribute and act with. Recognising common strategic threats and opportunities, such as those that globalisation, increasingly accessible big data sets and pervasive, or disruptive, technology may provide, can be key to creating and maintaining a successful Innovation strategy.

Before moving to the next chapter, where you will review your revenue and profit imperatives for Innovation, now may be a good time to ask yourself if you believe that your organisation has communicated the impact of these common imperatives for Innovation, and any others that you face, to all whom you wish to engage with in your pursuit of transformational or sustaining Innovation.

CHAPTER 3

REVENUE and CAPABILITY GAPS

Many organisations realise too late that they are trying to sell yesterday's products/services to today's customers. Sometimes they are fortunate and the decline is slow; sometimes more disruptive competitors arrive and effects are more dramatic. In either case declines in revenue, or worse, will likely occur if you find yourself in this position. To address this, wise innovators have to find ways to manage the delicate balance of investing in the future, whilst sustaining the benefits being obtained from their current core offerings.

To help quantify and subsequently act on such risks, wise innovators will consider using the commonly applied concept of assessment of 'Revenue Gaps', based on a simple analysis of an organisation's product and Innovation portfolio, to help focus needed specific activity or projects.

REVENUE (or expectation) GAPS

If you are visiting Innovation for the first time, then defining your profit or revenue gaps is a great place to start to focus your ambitions. Figure 3.1 is a simplistic representation of this concept as used by many wise innovators.

Figure 3.1 Using a Revenue Forecast to Identify Innovation Gaps

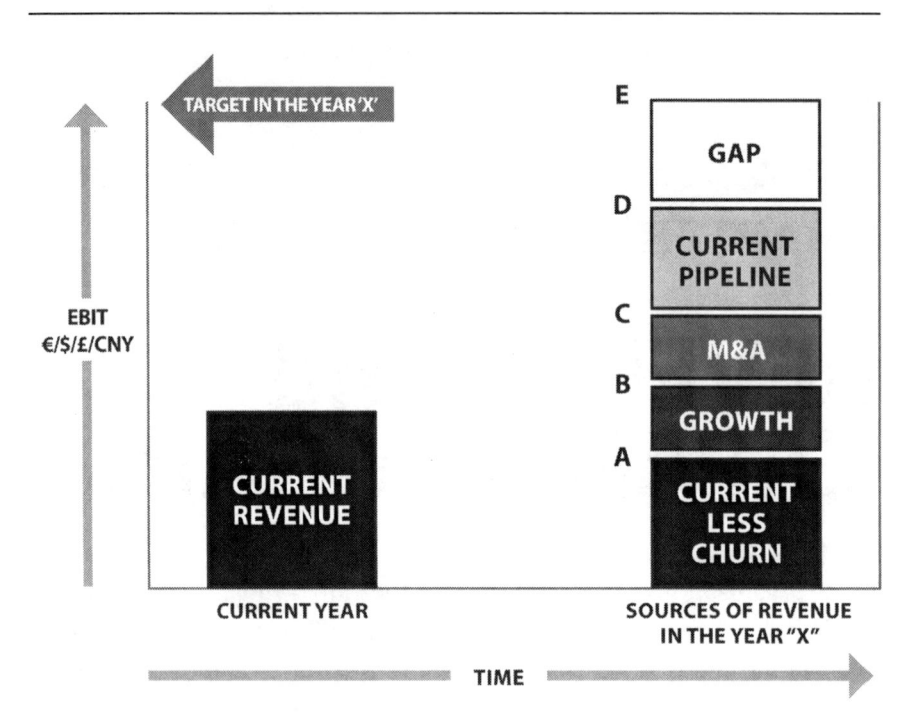

If you are in a public or non-profit sector you could also consider a similar analysis with the 'expectation gaps' of your Government, your Citizens or other key organisational stakeholders replacing EBIT (earnings before interest and tax) on the 'y' axis.

Wise innovators undertake analyses, such as these, to identify where desired future revenues will likely come from and to be able to identify, and act on, any gaps that may exist.

They will make predictions for revenue points 'A'–'E', in Figure 3.1:

'A': Revenue in the current year less any predicted loss from decline in market size or loss (churn) of current customers

'B': Adds an estimate of the growth that will be obtained in the current core business, for example from increases in total market size or new customers/market share

'C': Adds the effect of any planned mergers or acquisitions

'D': Adds a risk assessed view of the revenue likely to be generated from Innovations, currently in the Innovation project pipeline/portfolio, that will be launched in the period (See Chapters 7–9 of this book)

'E': Desired overall revenue in year 'x'.

Once this is done, any gaps (Revenue Target €/$/£/CNY at 'E' less revenue forecast a point 'D' in the simple example in Figure 3.1) are clear and a more detailed debate can take place about what Innovation projects should be initiated. This gap ('E'-'D') is your target for new innovation efforts. Specific Innovation ambitions, tactics and the sort of projects that may close such gaps are discussed further in Chapter 5.

CAPABILITY GAPS

Innovation abilities are not created 'overnight'. Wise innovators plan to develop, in advance of need, the competence (knowledge, expertise, etc.) and capacity (available resources to complete the task) that they predict they will need.

Wise Innovators will assess their baseline of Innovation-related capabilities and forecast, aligned to the activities that they foresee are going to be needed to close their revenue gaps, the future capabilities that they will need to develop, in order to be innovative and competitive in tomorrow's markets. How to do this is further described in the next chapter (Chapter 4).

CHAPTER SUMMARY

Clarity on revenue gaps (or a similar superordinate goal) can provide a useful focus around which to conclude Innovation imperatives and subsequently align Innovation strategies (the strategic impact of which will be further developed in Chapter 5). Clarity of ambition, alone, is never enough and wise innovators are continually evaluating their Innovation capabilities. Using an Innovation capability model may help such reviews and is further expanded on in Chapter 4.

Before moving to the next chapter, to learn more about Innovation Capability Assessment, now may be a good time to ask yourself if you have specific, clear, targeted, time bound, superordinate goal(s) for Innovation in your organisation and a way to review the capabilities and plans you will need to have in place to achieve them.

CHAPTER 4

CAPABILITY DEVELOPMENT and INNOVATION GOVERNANCE

Wise innovators will ensure that they have effective practices for Innovation governance. Using an Innovation Capability Assessment Framework can help managers to understand capability gaps and subsequently manage both specific Innovation programs and Innovation capability improvement plans, in both their own organisations and those with whom they may partner or collaborate with.

A wise innovator ensures that specific senior managers hold accountability for Innovation. This can usefully be linked to responsibilities for them to undertake assessments of their organisation's Innovation capabilities and achievements.

INNOVATION CAPABILITY ASSESSMENT

A key aim of this book is to provide a tool to enable readers to undertake Innovation capability reviews. The book is based on a 7-element Innovation Capability Assessment Framework that can be used to provide the basis for organisational capability and performance assessments. This framework is shown in overview in Figure 4.1 and fully explained in Chapters 5–11.

Figure 4.1 Innovation Capability Assessment Framework Overview

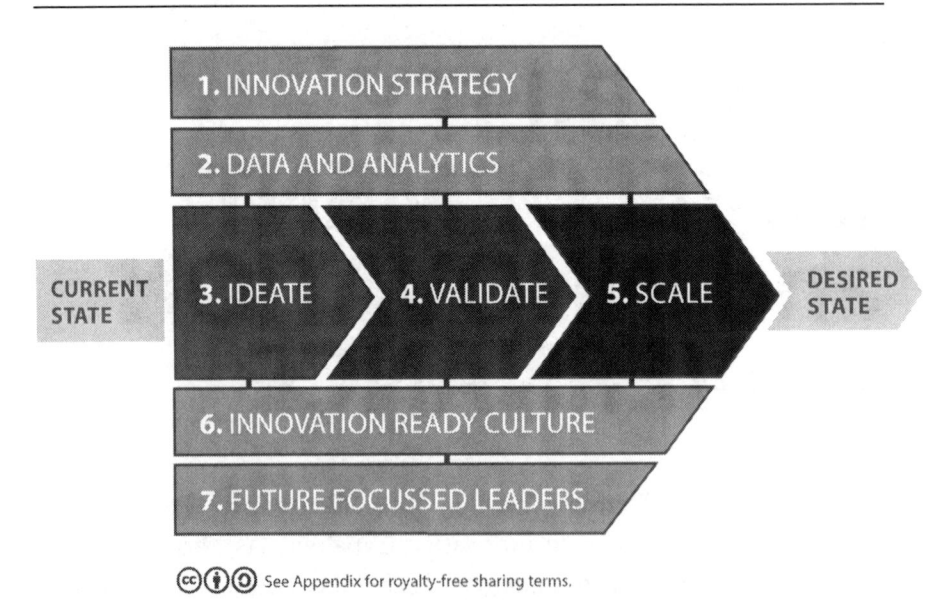

CURRENT STATE

1. INNOVATION STRATEGY
2. DATA AND ANALYTICS
3. IDEATE
4. VALIDATE
5. SCALE
6. INNOVATION READY CULTURE
7. FUTURE FOCUSSED LEADERS

DESIRED STATE

See Appendix for royalty-free sharing terms.

The logic of the Capability Assessment Framework is that a wise 'Innovation Strategy' (Element 1) will be informed by sound 'Data and Analytics' (Element 2), which will lead to a portfolio of projects in an ideas launch platform going from 'Ideation' (Element 3) to 'Validation' (Element 4) and 'Scaling' (Element 5), all of which is underpinned by 'Future

Focussed Leadership' behaviours (Element 7), which act to maintain an appropriate, supportive, 'Innovation Ready Culture' (Element 6). This logic is summarised in Figure 4.2.

Figure 4.2 Logic of the Innovation Capability Assessment Framework

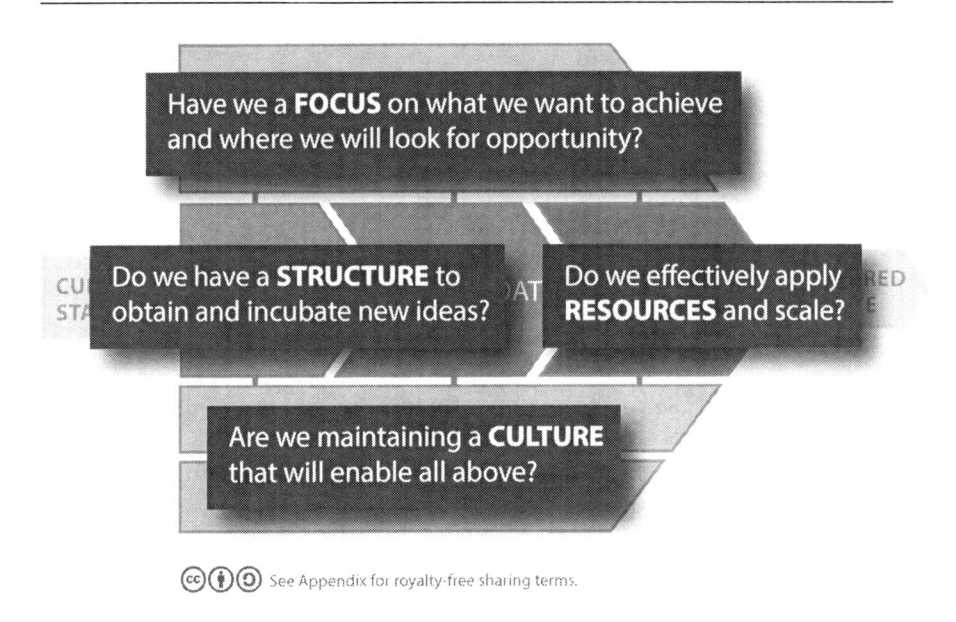

Have we a **FOCUS** on what we want to achieve and where we will look for opportunity?

Do we have a **STRUCTURE** to obtain and incubate new ideas?

Do we effectively apply **RESOURCES** and scale?

Are we maintaining a **CULTURE** that will enable all above?

See Appendix for royalty-free sharing terms.

Wise innovators undertake regular Innovation Capability Assessment following a process such as outlined in the eight steps below:

1. Select a steering team to undertake the assessment; typically, this would be a team of senior managers, often the Executive Management Team.

2. Brief the team and give them access to the questions of this workbook (the questions at the end of Chapters 5–11).

3. Consider whether a data-gathering phase will be beneficial and if so, whom to involve (e.g. other teams and managers).

4. Have all participants, individually, complete the analyses at the end of Chapters 5–11 and assign to specific individuals lead roles to facilitate subsequent reviews and discussions.

5. Undertake a workshop, with the aim of creating a consensual analysis and score (see scoring templates after the questions in Chapters 5–11).

6. For each improvement area identified, review the additional learning materials, ideas and good practice at www.innovationreboot.org.

7. Confirm or assign management responsibilities to those who will govern Innovation.

GOVERNANCE OF INNOVATION

Wise innovators know that the responsibility for Innovation should not reside with just one individual or team; a CEO, an Innovation department or Chief Technology Officer, alone, will not be likely to achieve enduring success without broader, company wide, engagement.

One way of achieving a broader based leadership of Innovation is to assign leadership governance roles across the elements of an Innovation Capability Framework. For example, within the Innovation Reboot Capability Framework in this book up to seven managers can be assigned governance responsibilities for overseeing capability development and achievements. Specific assignments could include:

Element 1 **Innovation Strategy:**
A manager accountable for overseeing the validity of your Innovation strategy, and the processes you use to conclude it.

Element 2 **Data and Analytics:**
A manager accountable for overseeing how well you scan and connect relevant knowledge.

Element 3 **Ideate:**
A manager accountable for overseeing the effectiveness and efficiency of how you create or acquire relevant ideas.

Element 4 **Validate:**
A manager accountable for overseeing how well you select likely 'winners' and maintain an innovation pipeline.

Element 5 **Scale:**
A manager accountable for overseeing the effectiveness and efficiency of how you scale and manage potentially successful Innovations.

Element 6 **Innovation Ready Culture:**
A manager accountable for overseeing if you are doing all you can to ensure that your overall business climate and culture encourages Innovation.

Element 7 **Future Focussed Leadership Development:**
A manager accountable for overseeing how you know if your leaders have the competencies to lead the business of tomorrow, as well as managing the organisation of today.

CHAPTER SUMMARY

In many organisations there is a gap between Innovation rhetoric (they want it) and reality (having agreed plans and actions on how to achieve it). Investing time in a systematic, regular review of Innovation capabilities and achievements can help overcome such issues. Using Innovation capability frameworks can not only help with assessment, but also provide a common language and structure within which to assign accountabilities and more actively, govern Innovation-related activities.

Before moving to the next section, ask yourself if you believe you have an appropriate, shared, consensual analysis of your organisation's Innovation capabilities/achievements and if you think you govern Innovation appropriate.

SECTION 2
CAPABILITY ASSESSMENT and MANAGEMENT

This section is structured around the seven elements of the Innovation Capability Assessment Framework. In Chapters 5 to 11 you can read insights into what 'wise innovators' do, with respect to each element, and then follow a series of questions to help you assess your own organisation's or team's successes with, and capabilities for, Innovation.

CHAPTER 5

⏻ ELEMENT 1: INNOVATION STRATEGY

In many organisations Innovation starts as an ad hoc collection of diverse activities, initiated by maverick managers and only occasionally systematically funded, when promising short-term gains are foreseen.

Organisations that have become successful at driving growth through Innovation know that such ad hoc approaches are not likely to deliver sustainable success. Wise and successful innovators address Innovation as a clearly defined strategic aim. They act on it, in planned and tangible ways, and in large organisations integrate related goals into their formal business plans, including targeting and resourcing of Innovation projects and Innovation capability building, over short-, middle- and long-term time horizons.

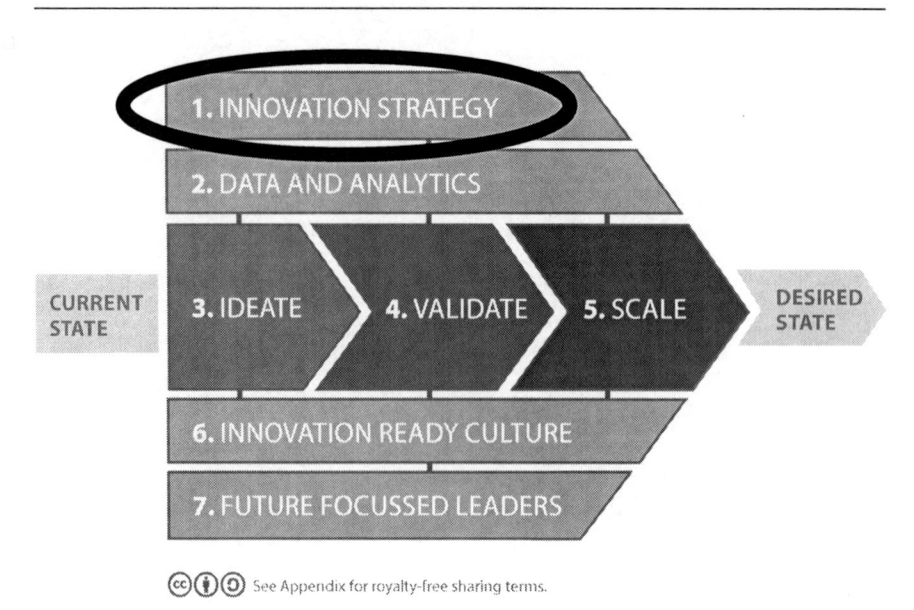

Figure 5.1 Innovation Capability Framework: Innovation Strategy

1. INNOVATION STRATEGY
2. DATA AND ANALYTICS
CURRENT STATE
3. IDEATE 4. VALIDATE 5. SCALE
DESIRED STATE
6. INNOVATION READY CULTURE
7. FUTURE FOCUSSED LEADERS

See Appendix for royalty-free sharing terms.

A wise innovator forecasts how much Innovation they need, what specifically it should achieve for them and, most importantly, in what areas and timeframes they should focus their budgets and efforts on.

KEY THINGS WISE INNOVATORS DO: **INNOVATION STRATEGY**

Wise innovators ensure that their strategic aims for Innovation are:

- Element 1.1: **Integrated to core**, i.e. aligned, tested and resourced in the same way as any other strategic aim or theme within the overall business strategy of their organisation.

- Element 1.2: **Based on focussed ambitions,** concluded after careful consideration of possible aspirations and risks, so as to create a realistic portfolio, some would say pipeline, of Innovation enabling projects that are spread over differing growth opportunity/risk categories and differing time frames.

- Element 1.3: **Systematically concluded** with assumptions, hypotheses and triggers recorded so they can be used to support ongoing management of subsequent Innovation projects or capability building.

ELEMENT 1.1:
INTEGRATION TO THE CORE BUSINESS STRATEGY

Most established organisations will have formal strategic processes by which they review the long-term purpose and identity of their organisation. If effective these process will articulate a desired, clear and inspiring statement of where the organisation needs/wishes to be positioned in the future, describe tactically how they intend to get there and articulate ways of working, processes and plans that guide others to deploy their concluded actions.

In a wise innovator, Innovation will almost invariably be seen as a key strategic theme and will be seamlessly integrated into such activities.

Common features of how this is done include:

- Having a clear transcendent vision showing the ambition they have for Innovation and how it contributes to achieving a desired future state for their organisation, their employees and society.

- Ensuring that related plans are capable of addressing their strategic Innovation imperatives and filling the revenue or capability gaps that exist.

- Being based on credible Data and Analytics (this is discussed in depth in Chapter 6).

- Spreading 'bets' with a balanced portfolio of projects designed to create a robust Innovation idea pipeline, with projects and ambitions in differing risk/aspiration categories.

- Assigning appropriate project resources, in budgets that are 'ring-fenced' (protected over multiple planning horizons) if they are for long-term Innovation projects.

Most of the points above are common good practice that will likely be considered well in any sound strategic process when Innovation is being taken seriously, the need for their attainment self evident.

It is, however, noteworthy that, in most wise innovators, the need to maintain a balanced portfolio of Innovation 'investments', based on realistic and focussed Innovation ambitions, is often perceived as being the key challenge.

ELEMENT 1.2:
FOCUSSED AMBITIONS

If you are not comfortable with risk, you will never be comfortable with Innovation. Any significant Innovation opportunity involves taking risk, rarely is it about just copying others. This means that strategic decisions need to be made *without* absolute confidence. There is no 'secret sauce' that is going to guarantee success, you are, after all, doing something that is likely to be new, new to you at least, so a history does not exist. You cannot get absolute 'proof' that what you are contemplating will work.

With such dilemmas in mind, wise innovators will use strategic tools to help them facilitate debates and make decisions on what Innovation projects/activities to invest in. Within large, established organisations a range of complex analytical tools may exist to assist them conclude and focus their innovation ambitions. Such tools are often based on the the work of H. Igor Ansoff[*1] and involve the use of an opportunity analysis matrix.

Figure 5.2 and the following examples show how the concept of an opportunity analysis matrix can be used, in any size of organisation, to help focus Innovation ambitions.

Figure 5.2 Opportunity Analysis Matrix –Two dimensions

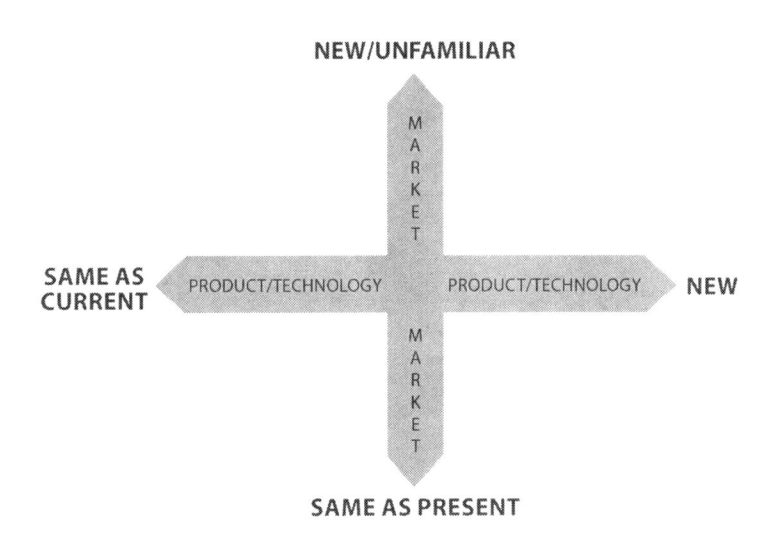

```
                    NEW/UNFAMILIAR
                          ▲
                          M
                          A
                          R
                          K
                          E
                          T
SAME AS    ◄── PRODUCT/TECHNOLOGY │ PRODUCT/TECHNOLOGY ──►   NEW
CURRENT                   M
                          A
                          R
                          K
                          E
                          T
                          ▼
                   SAME AS PRESENT
```

*[1]Ansoff, H. I., 'Strategies for Diversification', Harvard Business Review, Vol. 35 Issue 5, Sep-Oct 1957, pp. 113-124.

In the example above a matrix framework is ready to plot an organisations potential innovation investments based on an analysis of the organisation's presence/familiarity with or within a market or arena (the y-axis), against a scale for the evolution, or rate of change, of their products/technologies (the x-axis). It is then possible to plot potential Innovation projects into the four quadrants. The following Figure 5.3 shows an example of this.

Public sector users of this technique would align the y-axis to their familiarity with provision of a service to their stakeholders as opposed to the 'market' or 'arena'.

Figure 5.3 Opportunity Analysis Matrix –16 Potential Investments

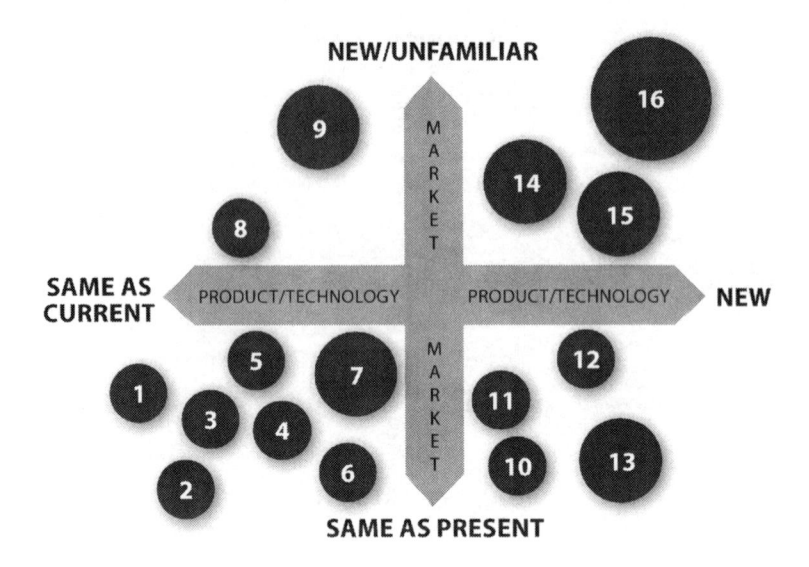

In the example given in Figure 5.3, potential Innovation investments are plotted with the circle size being indicative of the foreseen growth potential (e.g. growth in terms of financial revenue or value) of each of the potential projects. Each quadrant can now be seen to represent potential projects of differing strategic categories of Innovation.

Although there is no consensus on the terms that can be used to categorise the quadrants in such diagrams, most wise innovators will assign strategic categories to each quadrant. Terms such as 'Value Sustaining', 'Market Extension', 'Next Generation Product' and 'Transformational Growth' are often used.

Chapter 5: Element 1 **INNOVATION STRATEGY** 51

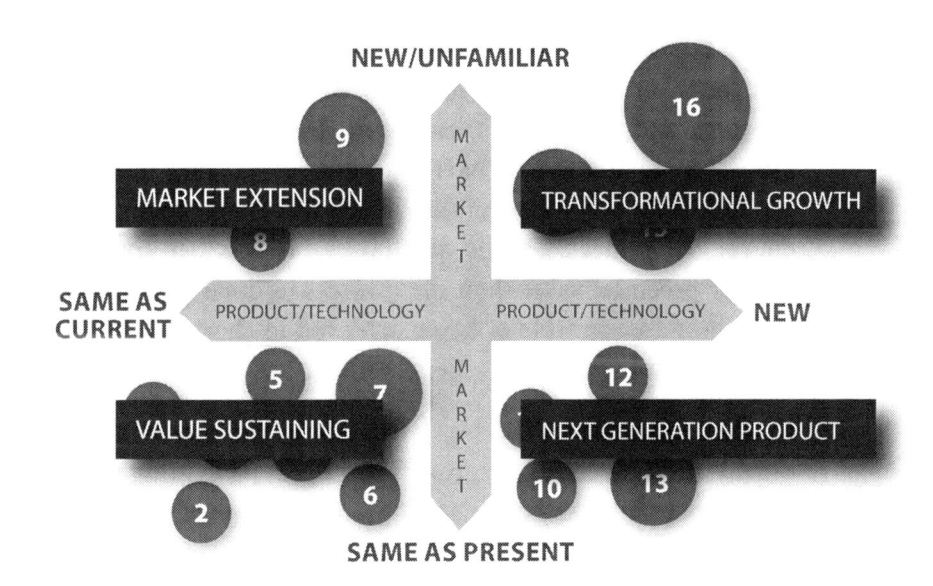

Figure 5.4 Opportunity Analysis Matrix – Strategic categories

---

[Diagram: Opportunity Analysis Matrix with two axes — vertical axis "MARKET" ranging from "SAME AS PRESENT" (bottom) to "NEW/UNFAMILIAR" (top), and horizontal axis "PRODUCT/TECHNOLOGY" ranging from "SAME AS CURRENT" (left) to "NEW" (right). Four quadrants:
- *Top-left: MARKET EXTENSION*
- *Top-right: TRANSFORMATIONAL GROWTH*
- *Bottom-left: VALUE SUSTAINING*
- *Bottom-right: NEXT GENERATION PRODUCT*

Numbered circles of varying sizes are distributed across the quadrants: 16 (top-right), 9 (top-left), 8, 5, 7, 2, 6 (bottom-left cluster), 12, 10, 13 (bottom-right).]

Value Sustaining Innovations: Lower value options (smaller circle size) will typically be in the bottom-left quadrant, and will likely be more numerous. These are about the sustaining of the performance of current products/services in current markets and are where the organisation will likely have sound, pre-existing market, technology and operational capabilities. Projects in this quadrant are often called 'Value Sustaining Innovations'.

Adjacencies for Market or Product Extensions: The next two adjacent quadrants can be seen as either 'Market Extensions' of predominately current products/technologies to new customers (top left), or 'Next Generation Products' obtained by product/technology extension to predominately existing customers (bottom-right).

Transformational Growth: The top-right quadrant is typically about new markets and significantly changed products/technologies, and is where Innovation opportunities may be big, but associated risks will be higher – as fewer pre-existing market, technology and operational capabilities will exist, as the market itself maybe totally new.

Each quadrant and category will have different growth opportunity features and benefits. In summary these are likely to be:

Value Sustaining: In the bottom-left are the investment options that focus on maintaining the value of an organisation's current offerings and protecting them from disruption from competitors. The organisation is likely to have pre-existing market and product/technology capabilities, which will enable them to easily deliver the projects in this quadrant. Projects here may be of smaller potential value than the projects in other quadrants, but they should be capable of being achieved in shorter time frames and with lower risks.

Market Extension: The top left is about the opportunities an organisation perceives for growth by migrating existing products or processes into new or adjacent markets. Think of the evolution of soft drinks like GlaxoSmithKline's Lucozade, that over time has moved from 'health' into 'sports' markets. Such projects can typically be scaled faster, and with less risk, than 'Transformational Growth' opportunities (top-right), as the organisation has some pre-existing product, if not market, capabilities. The risks and gains of these projects are typically higher than those in the 'Value Sustaining ' quadrant.

Next Generation Product/Technology: The bottom-right is about product extensions, where new products are evolved for predominantly existing customers/markets. Like the 'Market Extension' category, projects can typically be scaled faster and with less risk than 'Transformational Growth' opportunities (top-right), as pre-existing customer/market capabilities and probably some previous level of product/technology capabilities will exist.

Transformational Growth: The top-right is about targeting new markets with new products/services/models. This being commonly perceived as the toughest category in which to succeed as, unlike in 'Market Extension' or 'Next Generation Product' quadrants, neither market nor product/technology competence will likely

Chapter 5: Element 1 **INNOVATION STRATEGY** 53

pre-exist. Risks are higher, and time to market is typically longer. However, an element in any wise innovation strategy is to make some investments that attempt to create new space, push boundaries and achieve transformational growth.

Wise innovators spread their investment 'bets' across all quadrants and make their decisions based on quantification of differing levels of project risk and opportunity. Another way of looking at the same diagram is to analyse it as three levels of ambition/risk.

Figure 5.5 Opportunity Analysis Matrix – Level of Risk/Opportunity

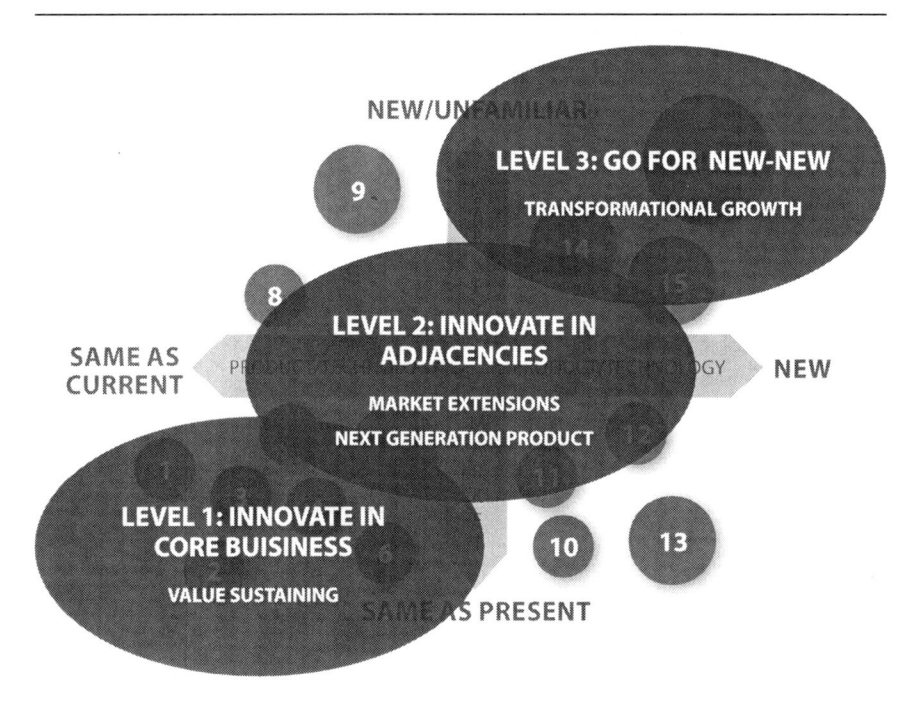

LEVEL 3: GO FOR NEW-NEW
TRANSFORMATIONAL GROWTH

LEVEL 2: INNOVATE IN ADJACENCIES
MARKET EXTENSIONS
NEXT GENERATION PRODUCT

LEVEL 1: INNOVATE IN CORE BUISINESS
VALUE SUSTAINING

NEW/UNFAMILIAR
SAME AS CURRENT
NEW
SAME AS PRESENT

The matrix analysis can now be seen as representing three levels of risk/opportunity, the key features of which are described overleaf:

Level 1: Innovate the Core Business – Investment 70%?

These are 'play not to loose' tactics. Activities at this level are typically directed towards reducing the risk of commoditisation of existing successful products/services, by activities such as creating variants to existing products and/or cost reductions. The achievements made will apply primarily to existing customers/markets and will typically be aimed at protecting existing markets, or incrementally increasing market shares.

Much is written[*2] about the risks of so-called 'low-end disruption', which occurs when a product of an established provider evolves too far, too fast, so that it disconnects from the need or rate at which established customers are prepared to buy/adopt new features. This creates an opportunity for a new entrant to join a market and to win sales based offering a lower cost product, typically lesser features. With risks such as this in mind, Innovation at this level is often as much about sustaining value, as it is about adding new features.

The amount of investment typically made in this category will vary dramatically, according to an organisations strategic need, but it is not uncommon for a mature high-performing company to use the often called "golden ratio" of innovation investments, in which 70 per cent of available resources are allocated at this level[*3].

[*2] Christensen, C. M., 1997 and 2011. 'The Innovators Dilemma', Boston, Harvard Business School Press and Harper Business reprint and Christensen, C. and Raynor, M., 2003. 'The Innovators Solution', Boston, Harvard Business School Press.

[*3] Nagji, B. and Tuff,G., May 2012, 'Managing Your Innovation Portfolio', Boston, Harvard Business Review. Reprint R1205C. with origins in Baghai, Mehrdad, Coley and White. 'The Alchemy of Growth: Practical Insights for Building the Enduring Enterprise'. New ed. Basic Books, 2000.

Projects in this category are usually capable of being undertaken in fairly short time scales and are unlikely to need radical changes to organisational structures, processes, resources or stakeholders. The risks are relatively low when compared to the following two levels.

Level 2: Innovate and Grow in Adjacencies – Investment 20%?

Investments here are about achieving growth by adjacent Product/Service growth (see bottom-right Figure 5.4 and 5.5) or Customer/Market (top left) extensions.

Projects initiated at this level are medium risk Innovation strategies that will build on some pre-existing successes with either current products and/or markets. Adjacent Innovation is often about leveraging something the company does well into a new space. As stated before, think of GlaxoSmithKline's Lucozade as a relevant example, also consider Nintendo, who's transition, a few years ago, with its Wii-fit platforms, involved taking its ability to make game consoles/interfaces and migrating it into a new health and fitness market.

The amount of investment will again vary according to an organisations strategic need. Many organisations will be seeking to achieve at least ten per cent of their overall revenue growth from expansion into adjacencies.

An often repeated ratio*[3] for Innovation investment at this level, in a mature, high-performing business, is for at least 20 per cent of the Innovation budget to be spent exploring potential projects here, but this maybe significantly more if more rapid growth is being sought, or technologies are changing rapidly.

There is an old saying that *'whilst an early bird may get the worm, it's the second mouse that gets the cheese'* and this applies to many Innovation successes. If you spot growth and Innovation opportunities that others are exploiting in adjacencies to you, then being a fast follower, by moving your business into markets or technologies that have been pioneered by others, can be an incredibly successful strategy (but note that slow followers are rarely as successful). Apple was not the first to sell mp3 players, when they first launched the iPod; it was to them, at that time, an adjacent market. They had the capabilities to respond quickly, they built a clever new ecosystem around their new product (iTunes), and they laid the foundations for what many respect as some of the most impressive growth attained in the last two decades, through a mixture of exploiting adjacencies for their technologies. It could also be argued that they also simultaneously created a transformational growth (see level 3 below) ecosystem in iTunes.

Level 3: Transformational Growth – Investment 10%?

Investment at this level is about 'playing to win big'. This is typically about both new markets and new technologies (new-new). Most organisations dream of creating the often called 'Blue Ocean' for transformational growth – uncontested market spaces about which much is written[*4], but these are typically high risk 'bets' and wise innovators do not pursue projects at this level in isolation of other strategies.

[*4] Kim, C. K. and Mauborgne, R., 2005. *'Blue Ocean Strategy: How to Create Uncontested Market Space and Make the Competition Irrelevant'*. Boston, Harvard Business School Press.

A route to a Level 3 'Blue Ocean' will often be navigated through adjacent (Level 2) steps where new capabilities can be incrementally built along the way. Projects at this level will often be large, complex and the opportunities are not likely to be frequent. Most successful innovators admit that, even if you are good at Innovation, you will probably only achieve one truly breakthrough Innovation, at this level, in a decade. The timescales are often long, the risks are invariably high and projects at this level need sustained investment commitments.

An often quoted example is Nike, who has undergone a fascinating transformation, with a range of both Level 2 and 3 Innovations, from its origin as a shoe company, into one where its future growth is now from digital technology (with phone apps, wearable devices, data and web services now a major part of its turnover). It has obtained growth from disruptive Innovation, but done so in a way that preserved the majority of their earlier markets.

A typical ratio for investment at this level[*3], in a mature, high-performing business, is for 10% of the Innovation budget to be spent here; but if higher growth rates are sought the figure will be significantly higher.

ELEMENT 1.3:
ASSUMPTIONS, HYPOTHESES and TRIGGERS

For many organisations history included times when success was all about finding a favourable position in a well-defined industry or market and then exploiting a long-term competitive advantage. For organisations today, such clear and sustainable competitive advantage is at best transient and for most a thing of the past.

Most organisations, large or small, public or private, need to maintain a range of ever changing hypotheses on "where to play" (arena's and market's), "how to win" (the value that will be offered) and the capabilities that will need to be developed along the way.

Wise innovators know that it is impossible to get all this 'right' all of the time and, recognising this dilemma, they 'plan to be reactive'; they log their strategic assumptions and hypotheses and continually seek insights and data to inform, confirm or dispel them.

They communicate their hypotheses and set related trigger/review. They are asking themselves not only *'What do I believe now?'*, but maybe more importantly, *'What would I have to continue to believe?'*, to be confident in the ongoing investment decisions being correct and against which background all subsequent ideation and validation thinking should be set.

ASSESS YOUR CAPABILITIES

It may now be timely to assess your own organisation's performance with respect to this element. A good way of doing this is to form an assessment team and provide them access to the following 10 questions, which are designed to help you debate and assess your current capabilities and achievements.

How to Assess:

1. Review whether a data-gathering phase will be beneficial, prior to answering the questions, and if so whom to involve.

2. Have each assessment team participant, individually, complete an analysis of each question, circling a score of 1 if they believe that few relevant outcomes/actions can be demonstrated, scoring 5 if they believe you truly have role model practices/outcomes and interpolating between scores 2 to 4 for other conclusions.

3. Undertake a workshop with the aim to create a consensual analysis and score for each question and debate what you think is the next big thing for you to work on.

What to do with the Assessment:

1. Take an average of your agreed scoring of the ten questions and conclude your current 'Reboot Status' for this element. Use this status number to form a base-line for future assessments and to share with interested stakeholders.

2. In areas that you choose to act on, seek insights and good practices that may help you conclude what to do next. We use funds from the sale of this book to fund the Innovation Reboot Project at www.innovationreboot.org. After considering what is the next big thing for you to address, take a look at this website for knowledge, ideas and discussions on each capability element.

Q1.1:
We are **clear on what we expect 'Innovation' to deliver.**

1. Unable to Demonstrate
2. In Parts
3. Basics Covered
4. Fully Able to Demonstrate
5. We're a Role Model

Q1.2:
Innovation projects are **appropriately resourced and planned.**

1. Unable to Demonstrate
2. In Parts
3. Basics Covered
4. Fully Able to Demonstrate
5. We're a Role Model

Q1.3:
Innovation plans **are translated in goals and communications** that having meaning for all involved.

1. Unable to Demonstrate
2. In Parts
3. Basics Covered
4. Fully Able to Demonstrate
5. We're a Role Model

Q1.4:
We have a **balanced portfolio of activities,** creating an Innovation pipeline that has a sound mix of projects, spread across a range of Innovation risk aspiration levels and timescales.

1. Unable to Demonstrate
2. In Parts
3. Basics Covered
4. Fully Able to Demonstrate
5. We're a Role Model

Q1.5:
Our Innovation plans **balance the tensions between short-term operational necessities and our Innovation goals.**

1. Unable to Demonstrate
2. In Parts
3. Basics Covered
4. Fully Able to Demonstrate
5. We're a Role Model

Chapter 5: Element 1 INNOVATION STRATEGY

Supporting Evidence:

Supporting Evidence:

Supporting Evidence:

Supporting Evidence:

Supporting Evidence:

Q1.6:
We always list **strategic assumptions and test our hypotheses.**

1	Unable to Demonstrate
2	In Parts
3	Basics Covered
4	Fully Able to Demonstrate
5	We're a Role Model

Q1.7:
We are effective at managing Innovation investments and setting related funding **trigger and review points.**

1	Unable to Demonstrate
2	In Parts
3	Basics Covered
4	Fully Able to Demonstrate
5	We're a Role Model

Q1.8:
We have **maintained our Innovation efforts even when times are tough** and short-term profits/savings are needed.

1	Unable to Demonstrate
2	In Parts
3	Basics Covered
4	Fully Able to Demonstrate
5	We're a Role Model

Q1.9:
Key **stakeholders have confidence in our Innovation plans**, our project 'pipeline', and are satisfied with the historic Innovation outcomes we have achieved.

1	Unable to Demonstrate
2	In Parts
3	Basics Covered
4	Fully Able to Demonstrate
5	We're a Role Model

Q1.10:
We have been able to ensure that **we do not experience disruptive or competitive Innovations** in our core business markets.

1	Unable to Demonstrate
2	In Parts
3	Basics Covered
4	Fully Able to Demonstrate
5	We're a Role Model

Chapter 5: Element 1 **INNOVATION STRATEGY**

Supporting Evidence:

Supporting Evidence:

Supporting Evidence:

Supporting Evidence:

Supporting Evidence:

CHAPTER 6

⏻ ELEMENT 2: DATA AND ANALYTICS

Successful innovators are often said to have been fortunate, to have 'seen the future before others', but such foresights are rarely good luck, they are more likely to be the outcome of the combination of thought through data acquisition plans and the use of sound analytical techniques that have helped them discover opportunity and make strategic and timely Innovation investments.

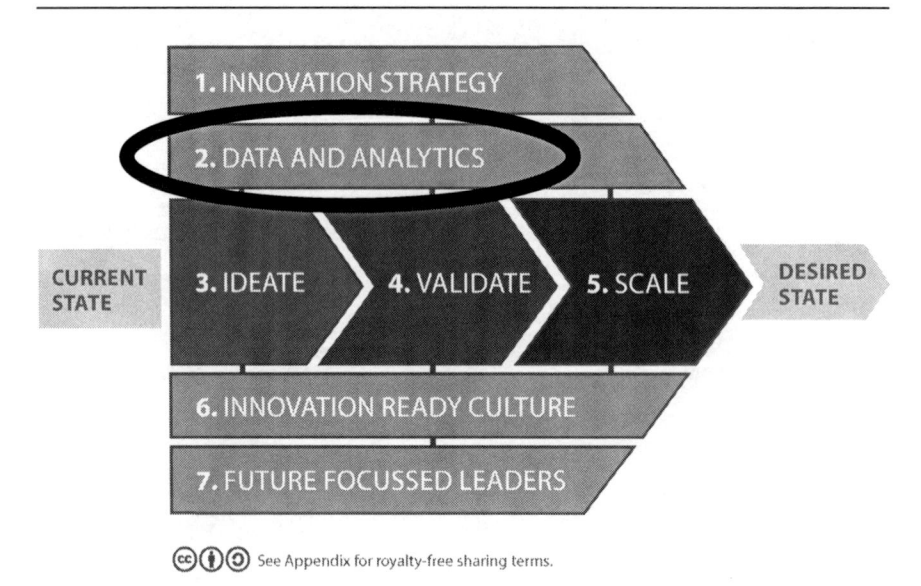

Figure 6.1 Innovation Capability Framework: Data and Analytics

1. INNOVATION STRATEGY
2. DATA AND ANALYTICS
CURRENT STATE
3. IDEATE
4. VALIDATE
5. SCALE
DESIRED STATE
6. INNOVATION READY CULTURE
7. FUTURE FOCUSSED LEADERS

See Appendix for royalty-free sharing terms.

The Innovation Capability Framework shows 'Data and Analytics' as a separate element, but this is not to suggest this necessitates having a discrete business function/team. This has been done in order to allow an Innovation-focussed data and analytics review to take place, within which the integration and effectiveness of all data collection and analysis activities, with Innovation in mind, can be evaluated.

KEY THINGS WISE INNOVATORS DO: **DATA AND ANALYTICS**

Wise innovators use data and analytics to support their Innovation planning. They will:

- Element 2.1: **Scan with broad horizons** to obtain, integrate and use a range of data sets and insights to enable them to spot opportunity or risk.

- Element 2.2: **Systematically analyse, connect and forecast** using valid techniques.

⏻ Element 2.3: **Take steps to avoid any decision-making biases** that may exist.

⏻ Element 2.4: **Systematically share emerging analyses** in order to engage others in broader debates and initiate idea generating activities.

ELEMENT 2.1:
BROAD HORIZONS – INPUTS

Wise innovators develop capabilities to listen to, integrate and make sense from, quantitative and qualitative data from relevant parts of their internal and external environments.

The cost of access to such data is declining. Vast quantities of data are generated every day and new data sources are frequently emerging. Additionally, mass market networked communication platforms provide opportunities for organisations to generate their own data streams. The ability to obtain and use such data is fast becoming itself, a basis of competition and growth.

Wise innovators know that it is focus (on quality not quantity), in data collection, that is key, as if the scope is too broad/deep it is possible to become defocussed, with signals that are unimportant and which will deplete available resources in evaluating them. Most wise innovators therefore spend time to review, segment and target where they will seek to obtain data and insight. Figure 6.2 illustrates some of the segments that can be considered.

Figure 6.2 Segmentation of Insight Opportunities

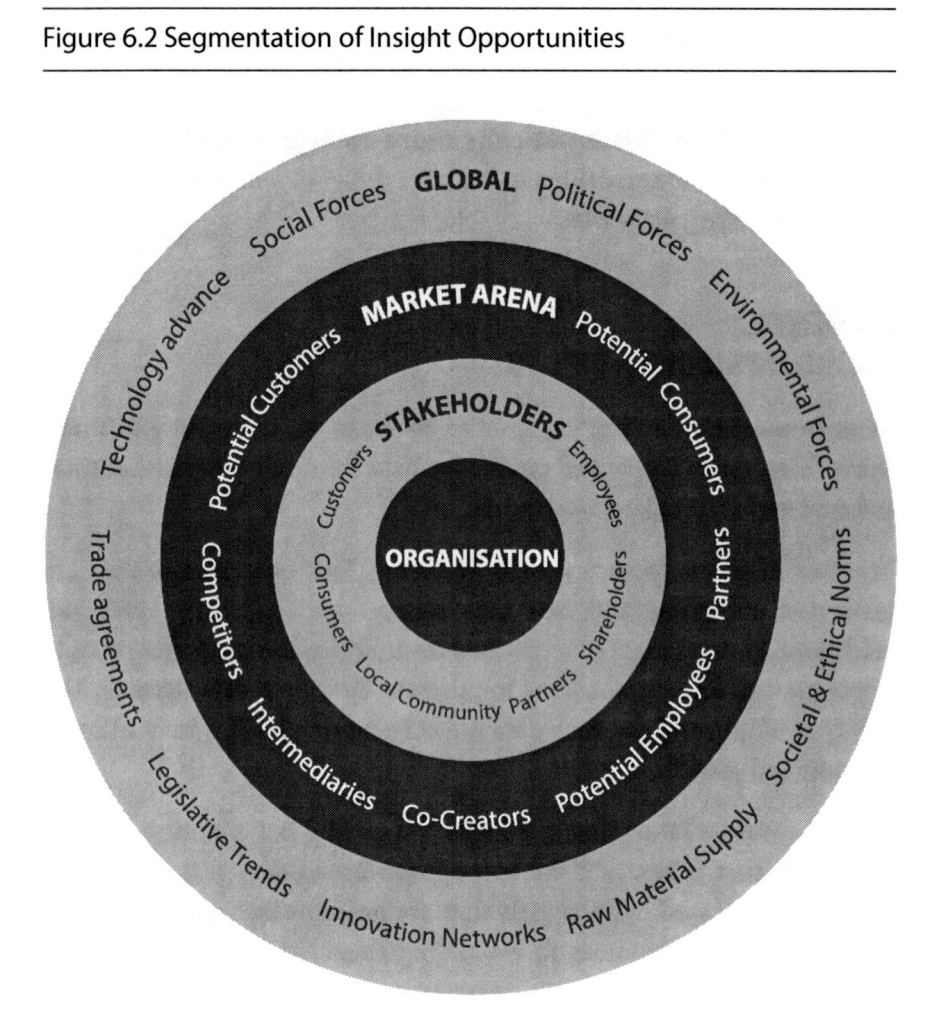

Wise Innovators will ensure that they have in place engagement networks (contacts, groups, etc.) or mechanisms (surveys, social media platforms, research, etc.) that can generate strategically targeted quantitative or qualitative data/insights in the following areas:

- The Global Strategic Environment:
  Including tracking political, legal, societal, environmental, economic and technological Forces (new technologies) that may have impact.

- Markets/Arenas and Competitors:
  Including tracking industry thought-leaders, media, competitors and complementors, potential customers, industry events and sociocultural trends in their market.

- Stakeholder's:
  Including obtaining empathetic understanding of, and deep insights from, their current customers (met and unmet needs, wants, etc.), consumers, intermediaries, distribution channels, partners, financial stakeholders and suppliers.

- The current Organisation:
  Including tracking the development of organisational competencies, mining the data from operations and using insights from technology experts supporting core business technologies (e.g. universities, research labs, Innovation networks, etc.).

Wise innovators will link their specific insight and data collection plans to their Innovation strategy; for example, if a significant part of an Innovation strategy is around being a 'Fast Follower' of competitors or markets, then related data and insight collection plans should have a focus on current competitors, adjacent potential competitors and general market signals.

If a strategy is based on exploiting an existing core competence, resource or technology that the organisation has built/acquired, then a wise organisation will want to closely scan for trends that may help them build on and/or protect their existing practices.

If disruption of an existing market is foreseen, then tracking trends of customer loss and the technologies being accessed/used by competitors maybe most key.

ELEMENT 2.2:
ANALYSING, CONNECTING and FORECASTING

Data is only useful if it can be used to drive outcomes. By itself, just collecting and integrating data does not generate value. Wise innovators develop skills that make them able to make ever-faster sense, meaning and opportunities out of the flood of signals they receive. Innovation comes from not only having the data and insights, but being able to connect and act on them to create meaningful foresights that may help to predict strategically important aspects for the future, such as the demands and tastes of customers, the vectors of technologies and the actions of competitors.

Increasingly, achieving this means obtaining access to analytical capabilities to filter the irrelevant, hypothesise patterns or connections, and support managers in interpreting what they have obtained. Wise innovators are continually trying to add value by using analytical techniques that 'distil' the wisdom from data, in a way that provides them with actionable foresights. They are continually trying to travel as far down the road from 'data' to 'foresights' as they can get.

Figure 6.3 The 'Road' from Data to Foresight is Long

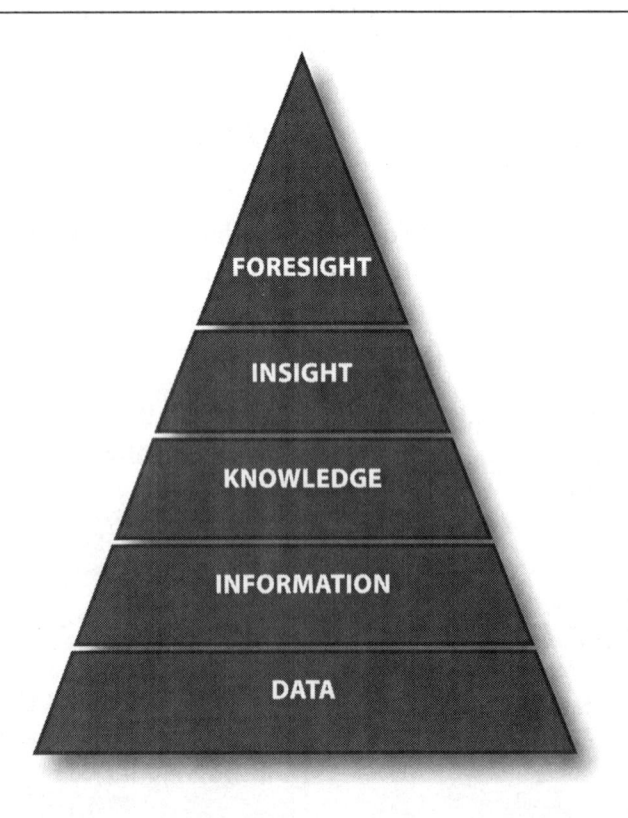

Data: Observations or facts in raw or unorganised form, if used well, can lead to **Information:** Raw data that has been analysed and organised, for a purpose and presented in a way that gives it meaning and relevance, which can then lead to **Knowledge:** An understanding or belief about a subject which is justified by supportive alignment of accumulated data, information and experiences, can in turn lead to an **Insight:** An understanding of the forces behind an action or its outcome, which can lead to the ultimate aim of producing **Foresights:** The ability to detect/predict significant upcoming change in a way that enables the prediction to be used in long-term decision-making.

Most wise innovators respect the wisdom of Peter Drucker[*1] as summarised in his classic book *'Innovation and Entrepreneurship'*, in which he argues that *'major innovations are almost never based on the extrapolation of one factor, but on the convergence of several different kinds of knowledge'*. This insight, that innovation typically occurs at the intersections of bodies of knowledge, has endured the test of time.

[*1] Drucker, P. F. *'Innovation and Entrepreneurship'*. 2Rev Ed. A Butterworth-Heinemann Title, 2007.

Wise innovators will seek to 'feed' themselves the right data and information 'diet', and then 'distil', 'ferment' and 'mix' their data to find connections, convergences, insights and foresights that can help them make decisions with more certainty and needing less 'faith'.

To help with the journey (distilling data into insights and foresights), most large and wise innovators use intelligent systems to help them find connections within and between bodies of knowledge. Analysis and futuring techniques like: quantitative data analysis, predictive modelling, data mining, decision-tree mapping, Delphi, and many similar terms, will trip off the tongue of large and successful innovators, as will the names of providers of analytical software and databases that may assist them.

Small organisations and entrepreneurs, however, should note that easy access to such analytical capabilities are no longer economically restricted to large companies, with large budgets. Ever-cheaper computing power and an increasing number of 'no-cost' and open source cloud platform computing solutions (such as Apache Hadoop), are making it economically feasible for anyone to obtain and analyse large quantities of data and seek connections. Wise innovators, be they large or small organisations, are realising that if they do not embrace 'mining' data for connections and opportunities they are at risk from better informed competitors who do.

To help with spotting connections, wise innovators, large and small, will often assign 'trend owners' to key data sets, i.e. managers who have accountability to investigate and make trend extrapolations on key data.

As Drucker observed, most find that Innovation opportunities occur at the intersection of trends; for example, the social trend of globally ageing populations is currently intersecting with trends and advances in technologies that allow Innovations in home medical diagnosis and treatment techniques. Wise innovators ensure that their trend owners meet regularly and have management processes to help them identify patterns, points of trend intersections and inflexions that may help them identify potential opportunities.

ELEMENT 2.3:
AVOIDING DECISION-MAKING BIAS

Management debates and decisions always involve a mixture of intuition, beliefs and facts. How a person mentally handles their personal beliefs and how group dynamics are managed in team discussions, will affect the validity of decisions made. Wise innovators recognise that it is very easy for a weak signal of opportunity, or an embryonic idea, to be supressed/missed, and are sensitive to risks in areas such as:

- Poor Group Dynamics
  They will ensure that meetings are facilitated, to help overcome any potential bias caused by the most vocal or most senior persons' views, leading to a narrow-minded analysis and a false 'consensus'.

- Confirmation Bias
  They will challenge those they perceive as seeing (or seeking) what they expect, rather than understanding the reality.

- Political Bias
  They will challenge those who will not say what needs to be said, or are unwilling to challenge the status quo.

- Confused Correlations
  They are mindful to ensure that patterns are analysed well and that they do not confuse coincidence, correlation and cause. For example, up to the age of 16, larger shoe sizes correlate with better reading ability, but enlarged feet are not the cause of better reading ability. The causal variable is brain size/age. Spotting a correlation does not in itself prove causation. Wise innovators know that scientific rigor needs to be applied to test causality.

ELEMENT 2.4:
SYSTEMATICALLY SHARE EMERGING ANALYSES

Wise innovators share information with a passion and a purpose. Information, knowledge, insights and foresights only have meaning when they reside in people's minds and so need to be shared with those who can comment, contribute and act on them.

Whether by using the Innovation modules within enterprise-wide software systems (increasingly common in large organisations), or simply summarised in other ways, 'opportunity or challenge briefs' (Figure 6.4 box 'A') are often used to document the areas in which opportunities have been found and in which ideas are sought. In wise innovators such opportunity briefs will summarise all available and relevant data, information, knowledge, insights and foresights in order to frame the ways in which the organisation will begin to engage those who may be able to generate useful ideas, to be pursued in the Ideate/Validate/Scale pipeline in elements 3-5 of the Innovation Capability Framework (and as described in detail in Chapters 7–9).

Figure 6.4 Create Opportunity Briefs in Areas of Strategic Interest
---

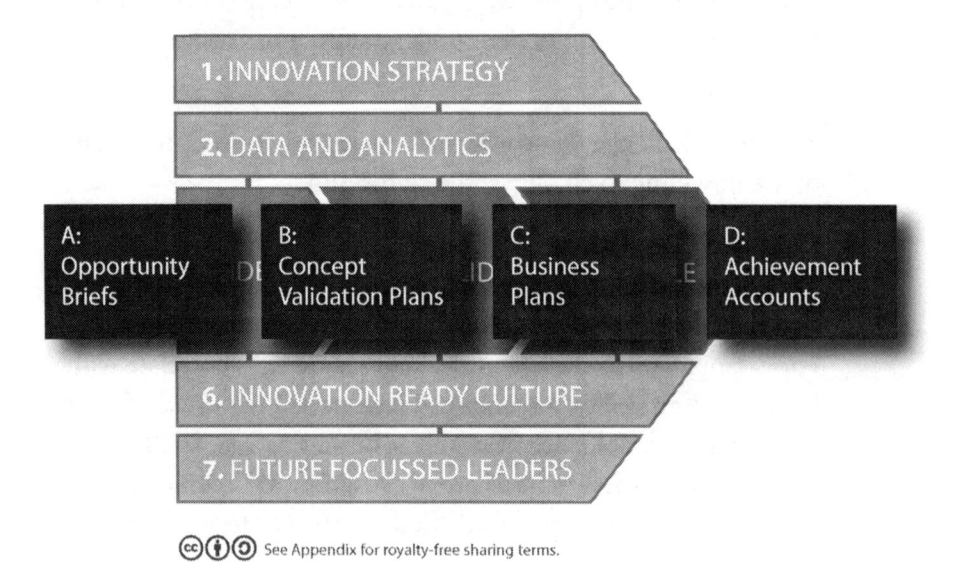

1. INNOVATION STRATEGY

2. DATA AND ANALYTICS

| A: Opportunity Briefs | B: Concept Validation Plans | C: Business Plans | D: Achievement Accounts |

6. INNOVATION READY CULTURE

7. FUTURE FOCUSSED LEADERS

See Appendix for royalty-free sharing terms.

A well structured Opportunity Brief (Figure 6.4 box 'A') will answer questions like *(fictitious, large project, example shown in the italics below)*:

- What is the core idea on which ideas are sought?
  - Making an innovative new input interface that is simpler and faster...

- What is the strategic opportunity?
  - Our software can run faster than the current interface; so if we can improve the interface we can leapfrog our competitors in an area that customers have expressed as a desire. See infographic 'x' and 'y'. It fits with our strategy to achieve growth in...

- What insights/foresights/hypotheses have we concluded?
  - Technology foresights lead us to predict that significant advances to tactile touch technology and head-up display technologies are just over the horizon. Social trends tell us they are seen as cool...

- What is the profile of the potential customers?
  - All current customers. We particularly believe that under 25's will see this as more attractive than our competitors offerings, but this is as yet not validated...

- Where will we focus idea/input seeking?
  - The Cambridge University team on tactile touch technologies + Open call on the intranet to all our employees, particularly, for input on the physical shape + structured co-creation debate with lead supplier 1, along with...

- What are the time frames and exploratory budgets?
  - These are...

- What features maybe desirable in potential solutions?
  - Must be light, capable of screen printing, capable of design protection, etc...

ASSESS YOUR CAPABILITIES

It may now be timely to assess your own organisation's performance with respect to this element. A good way of doing this is to form an assessment team and provide them access to the following 10 questions, which are designed to help you debate and assess your current capabilities and achievements.

How to Assess:

1. Review whether a data-gathering phase will be beneficial, prior to answering the questions, and if so whom to involve.

2. Have each assessment team participant, individually, complete an analysis of each question, circling a score of 1 if they believe that few relevant outcomes/actions can be demonstrated, scoring 5 if they believe you truly have role model practices/outcomes and interpolating between scores 2 to 4 for other conclusions.

3. Undertake a workshop with the aim to create a consensual analysis and score for each question and debate what you think is the next big thing for you to work on.

What to do with the Assessment:

1. Take an average of your agreed scoring of the ten questions and conclude your current 'Reboot Status' for this element. Use this status number to form a base-line for future assessments and to share with interested stakeholders.

2. In areas that you choose to act on, seek insights and good practices that may help you conclude what to do next. We use funds from the sale of this book to fund the Innovation Reboot Project at www.innovationreboot.org. After considering what is the next big thing for you to address, take a look at this website for knowledge, ideas and discussions on each capability element.

Q2.1:
We maintain a **good 'peripheral' vision** and monitor relevant data that may lead to valid strategic insights.

1	Unable to Demonstrate
2	In Parts
3	Basics Covered
4	Fully Able to Demonstrate
5	We're a Role Model

Q2.2:
We do not suffer from 'data silos'. **We are effective at integrating data and insights** from multiple sources.

1	Unable to Demonstrate
2	In Parts
3	Basics Covered
4	Fully Able to Demonstrate
5	We're a Role Model

Q2.3:
We use a valid **relevant range of analytical techniques** to help us connect data, information, knowledge and insights.

1	Unable to Demonstrate
2	In Parts
3	Basics Covered
4	Fully Able to Demonstrate
5	We're a Role Model

Q2.4:
We have **accountability assigned for investigating key data sets/emergent trends**.

1	Unable to Demonstrate
2	In Parts
3	Basics Covered
4	Fully Able to Demonstrate
5	We're a Role Model

Q2.5:
We **keep and share trend, assumption, event triggers and decision records** across the organisation.

1	Unable to Demonstrate
2	In Parts
3	Basics Covered
4	Fully Able to Demonstrate
5	We're a Role Model

Supporting Evidence:

Supporting Evidence:

Supporting Evidence:

Supporting Evidence:

Supporting Evidence:

Q2.6:
We understand the risks of **analysis and decision bias**, our support and facilitation tools help us minimise them.

1. Unable to Demonstrate
2. In Parts
3. Basics Covered
4. Fully Able to Demonstrate
5. We're a Role Model

Q2.7:
We are **prepared to challenge the fundamental assumptions that underlie our strategies and plans.**

1. Unable to Demonstrate
2. In Parts
3. Basics Covered
4. Fully Able to Demonstrate
5. We're a Role Model

Q2.8:
We are good at **communicating Innovation opportunity briefs** to those whom are able to help us ideate.

1. Unable to Demonstrate
2. In Parts
3. Basics Covered
4. Fully Able to Demonstrate
5. We're a Role Model

Q2.9:
We have achieved a **better understanding of our current customers and end-users than our competitors** have.

1. Unable to Demonstrate
2. In Parts
3. Basics Covered
4. Fully Able to Demonstrate
5. We're a Role Model

Q2.10:
We can demonstrate that we are **identifying and acting on emerging trends ahead of the 'pack'.**

1. Unable to Demonstrate
2. In Parts
3. Basics Covered
4. Fully Able to Demonstrate
5. We're a Role Model

Chapter 6: Element 2 **DATA and ANALYTICS**

Supporting Evidence:

Supporting Evidence:

Supporting Evidence:

Supporting Evidence:

Supporting Evidence:

CHAPTER 7

⏻ ELEMENT 3: IDEATE – CREATE OR ACQUIRE IDEAS

Wise innovators know how to ideate. They know that however good that their Innovation strategy is, and however robust their data and analytics are, without a supply of new ideas little progression will be made.

They recognise that a capability to ideate is a core skill of an innovative organisation and that both internal and external sources of ideas need to be maintained.

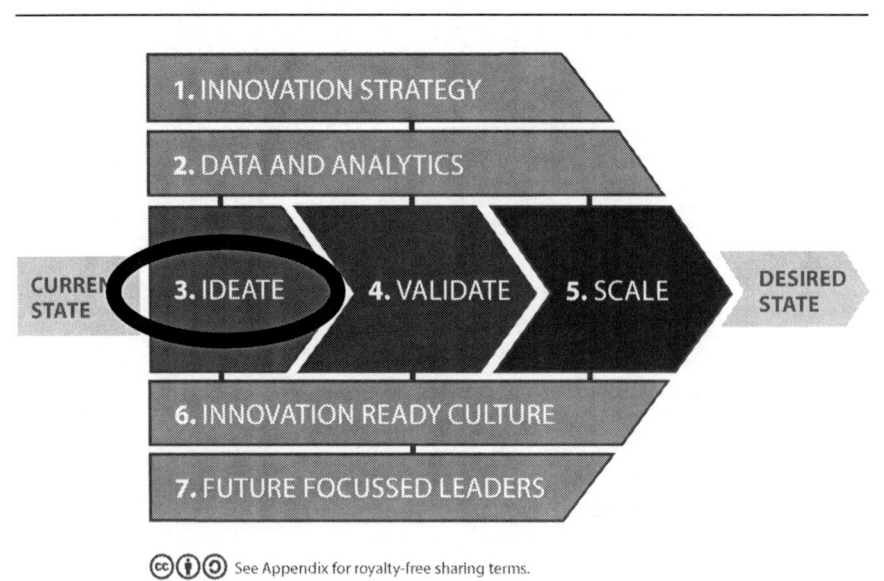

Figure 7.1 Innovation Capability Framework: Ideation

1. INNOVATION STRATEGY
2. DATA AND ANALYTICS
3. IDEATE
4. VALIDATE
5. SCALE
6. INNOVATION READY CULTURE
7. FUTURE FOCUSSED LEADERS

CURRENT STATE → DESIRED STATE

See Appendix for royalty-free sharing terms.

In wise innovators, all activities that are linked to the Ideate>Validate>Scale central part of the Innovation capability model are systematically managed; they are not ad hoc or unfocussed activities. Effective ideation and subsequent Innovation are rarely achieved by 'good luck'.

KEY THINGS WISE INNOVATORS DO: **IDEATE**

Wise innovators build capabilities to ideate. They are able to:

- Element 3.1: **Obtain Ideas from inside the organisation,** using a range of techniques to maintain a stream of internal creativity and contributions.

- Element 3.2: **Use external sources of ideas,** knowing when and how to cultivate and integrate ideas from outside.

- Element 3.3: **Use idea management systems** to systematically develop, record and share ideas, for use both now, and in the future.

⏻ Element 3.4: **Systematically select likely winners,** by holding strategically focussed reviews of their ideation outcomes and selecting a range of the most promising ideas, for further validation.

ELEMENT 3.1:
CREATIVITY FROM INSIDE THE ORGANISATION

Great ideas can come from 'inside' an organisation, but this is not likely to happen unless a climate in which ideas are sought and welcomed has been created. Maintaining effective internal ideation capability takes time and requires investment in resources and processes to support it.

Wise innovators will consider using a range of techniques such as:

- Having Innovation/creativity champions to facilitate ideas generation.

- Maintaining subject matter experts and domain expertise in targeted technical or managerial subject areas, which are likely to be useful to support their Innovation aims. *For example, in technology companies, the technical domain of 'interface technology' is likely to be an area where subject expertise and internal communities of practice will be useful to stimulate debate and ideas.*

- Broadcasting success stories, technology and competitor advancements, in order to increase awareness and stimulate debate and thinking.

- Giving employees time to allow 'thinking' and exploration of ideas.

- Using a range of thinking tools, forums and processes to stimulate ideas. For example, small-group brainstorming, prototyping, real-world observational analysis and such like.

- Implementing suggestion schemes or holding idea jams or contests.

- Emphasising and respecting diversity and understanding that diverse teams/forums can often bring new perspectives to established thinking.

Figure 7.2 Wise Innovators use a Range of Techniques

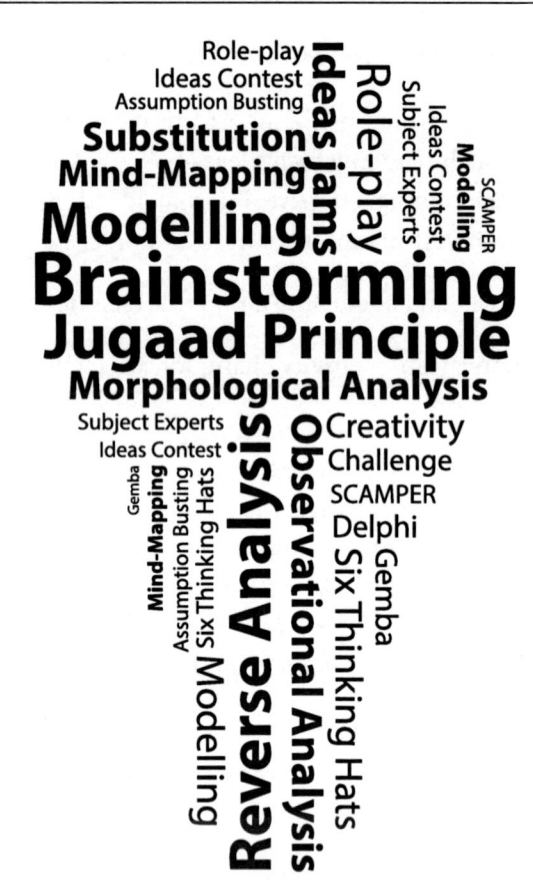

Role-play, Ideas Contest, Assumption Busting, Substitution, Ideas jams, Mind-Mapping, Role-play, Ideas Contest, Subject Experts, Modelling, SCAMPER, Modelling, Brainstorming, Jugaad Principle, Morphological Analysis, Subject Experts, Ideas Contest, Creativity, Challenge, SCAMPER, Delphi, Gemba, Mind-Mapping, Assumption Busting, Six Thinking Hats, Reverse Analysis, Observational Analysis, Modelling, Six Thinking Hats, Gemba

Generating ideas from inside has many benefits, not least, that the organisation has total control over how to develop any resultant good ideas and can own the intellectual property of the resultant outcomes.

However, internal ideas generation can come with the risk of being too incremental, internally focussed and slow (external ideation can often be faster). There are also internal challenges that may need to be addressed before it can be effective, particularly in an established organisation, where cultural issues may need to be overcome (Chapter 10: 'Innovation Ready Culture' expands on how wise innovators address such issues).

ELEMENT 3.2:
OBTAINING IDEAS FROM OUTSIDE

Most organisations will experience a need to look outside to find non-core expertise. Wise innovators also recognise that diversity is often key for truly creative thinking and that external ideation can help expand potentially constrained internal thinking. The key question(s) that most successful innovators debate is not whether external sources are valid, but which ones should be systematically used, plus when and how should they be managed.

Wise innovators are able to work with external inputs to the innovation process, as easily as they do with internal inputs. They will consider:

- Working collaboratively with partners, customers, suppliers, vendors, and even competitors.
- Contract research projects, or participating in research consortia.
- Joint ventures with risk- and reward-sharing.
- Creating 'open', web-enabled portals, or engagement models, to engage 'crowds' of skill and Innovation capacity quickly.
- Externally focussed ideas contests.
- Bringing relevant parties together in Innovation-focussed events (often called 'Innovation Jams').
- Use of consultants/ideas scouts.
- Use of intermediary idea, R&D or intelligence platforms (such as Innocentive, Nine Sigma, openIDEO, Kaggle, et al).

- Using social networking technologies to draw consumers (and others) into specific product development processes and inform internal teams.

- Acquisitions of companies with needed competencies, ideas or Intellectual Property (IP).

External ideation partnerships bring with them their own cost and cultural issues. Overcoming the 'not invented here, won't work for us' syndrome takes time, as does managing the external interfaces created and maintaining appropriate IP rights and confidentiality.

ELEMENT 3.3:
IDEAS MANAGEMENT

Whatever the source of ideas (internal or external), wise innovators invest time to create a 'corporate memory bank', that helps them systematically debate, review and register ideas. They know that a well implemented ideas management system will help timely administration of ideas, help identify past work, which may now be of use and ensure that resources are not wasted on ideas that have been previously rejected, for reasons that are still valid.

Figure 7.3 Wise Innovators Keep Ideas Registers

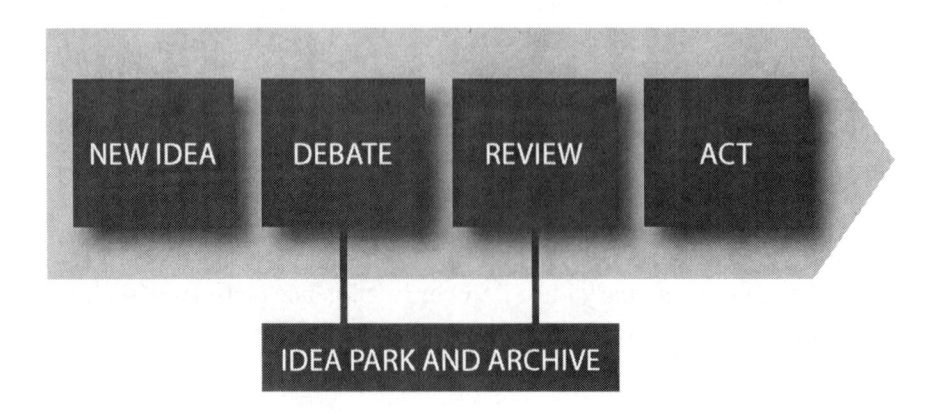

NEW IDEA → DEBATE → REVIEW → ACT

IDEA PARK AND ARCHIVE

Whether in enterprise-wide systems or simpler ad hoc processes, wise innovators will create a portal for idea submissions, alignment, review and tracking. They will allow as open an access as possible (bounded only by confidentiality and IP risk issues) to the portal and encourage dialogue, on going sharing and capture of new knowledge. Commonly available idea/Innovation systems (such as IdeaScale, INova, Spigit and many others), also bring with them an ability to act as community platforms for sharing and evaluating ideas and for project managing the outcomes.

Smaller organisations can often make good use of existing intranets/enterprise management systems, for ideas management and make use of freely available social technologies (Facebook, Twitter, etc.), as platforms for wider engagements/sharing.

ELEMENT 3.4:
SELECTING LIKELY 'WINNERS' TO FURTHER VALIDATE

Most wise innovators will argue that *'you do not want a 'creative' company, a company that is just full of ideas, but you do want an 'innovative' company, a company that can select the best/valid ideas and then put them into practice'*. Once ideas are flowing, a most critical next step is that of combining and, then selecting, the most promising, potentially 'winning' ideas and creating validation plans to develop and test the ideas and concepts that look likely to be of value (Figure 7.6 –B).

Figure 7.4 Create Validation Plans

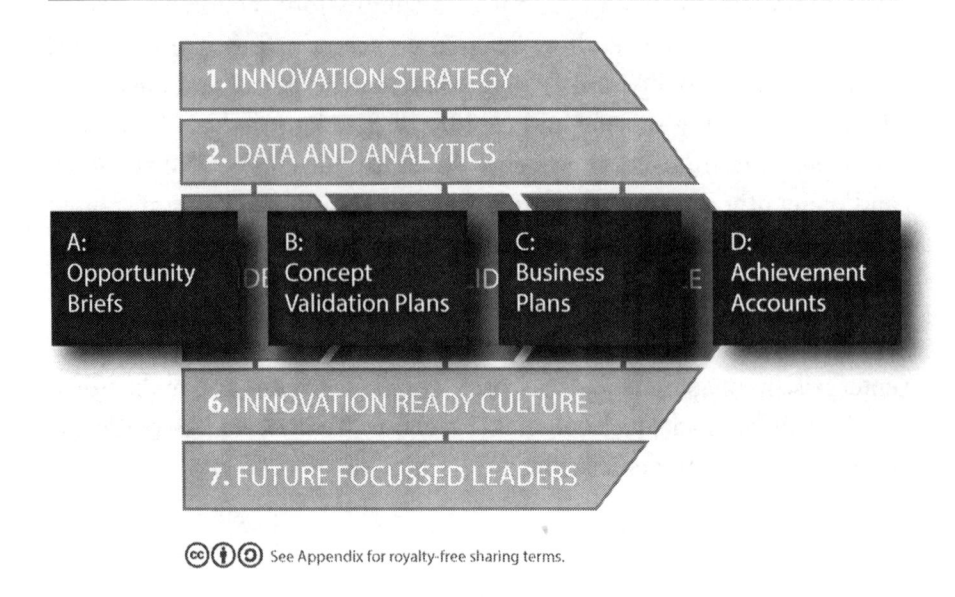

```
1. INNOVATION STRATEGY
2. DATA AND ANALYTICS

A:                B:              C:              D:
Opportunity       Concept         Business        Achievement
Briefs            Validation      Plans           Accounts
                  Plans

6. INNOVATION READY CULTURE
7. FUTURE FOCUSSED LEADERS
```

See Appendix for royalty-free sharing terms.

Ideas selection can become a big task as the number of ideas generated can be very large, particularly, if care has not been taken to produce clear challenge or opportunity briefs, that focus where inputs are being sought.

The ratio of raw ideas to products/services launched will vary dramatically, according to the nature of an innovator's core business, but to give an (potentially high end) example, Figure 7.5 summarises numbers, based on the work of Stevens and Burley at Dow Chemicals[*1], that illustrate a top-end view of what could arise from general or unfocussed ideation campaigns.

Figure 7.5 Wise Innovators know their Idea Conversion Ratios

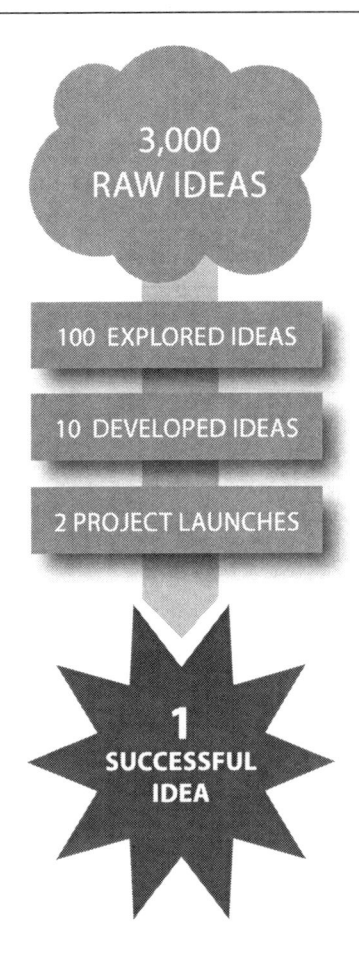

- 3,000 RAW IDEAS
- 100 EXPLORED IDEAS
- 10 DEVELOPED IDEAS
- 2 PROJECT LAUNCHES
- 1 SUCCESSFUL IDEA

[1] Stevens, G. A. and Burley, J., 1997. '3000 Raw Ideas = 1 Commercial Success!' Research-Technology Management 40(3), pp.16-27.

Whether in volumes such as in Figure 7.5, or in more focussed campaigns, idea filtering (being able to group, prioritise and develop ideas) is a key skill that wise innovators develop. Wise innovators will put in place systematic solutions to vetting and selecting 'winners', in order to ensure they are selecting the most promising and strategically relevant ideas. They will also ensure that the decisions made are transparent and well communicated, so as to maintain the motivation of all involved, particularly, those whose ideas were not selected for further validation.

For large organisations filtering solutions are often present in enterprise-wide Innovation Management software and can be amplified and/or replaced by vetting processes such as crowd-vetting (transparent voting competitions), internal ideas markets (trading with fictitious currency), or just standard strategic prioritisation tools, such as ease-impact matrices. Small organisations may find a simple ease–impact matrix of particular use.

Figure 7.6 Idea Evaluations in a Simple, Ease–Impact Matrix

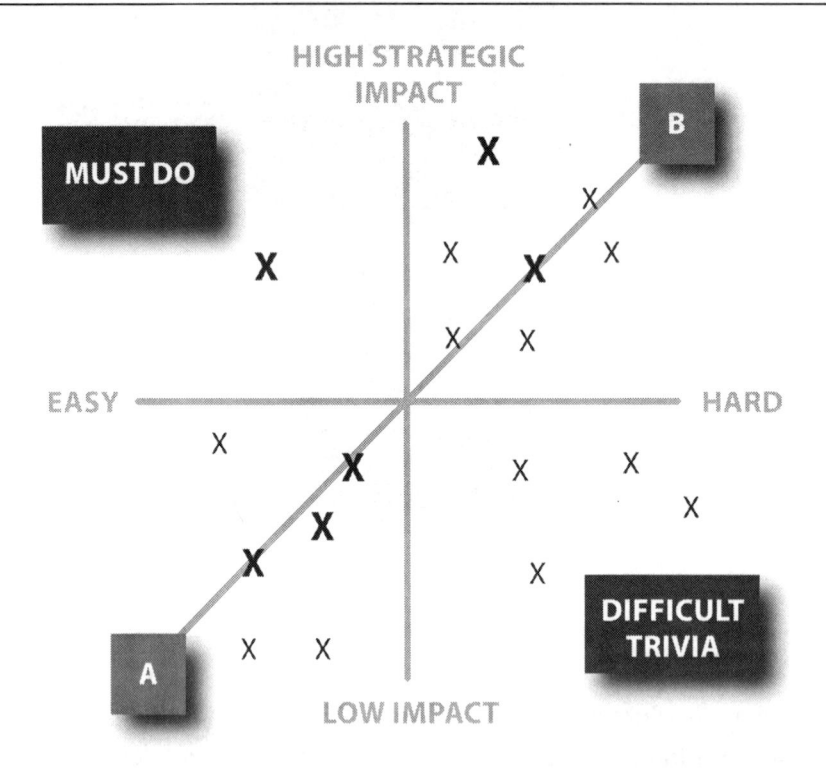

---

y-axis – The potential impact of each idea (its likely contribution to the organisation's specific Innovation strategy or a specific 'opportunity brief' as described in Chapter 6.)

x-axis – The foreseen difficulty in doing it (time, money, effort, speed, fit to existing products, etc.).

How to use: After taking the easy decisions (top left and bottom-right) most debates will resolve around the line A to B. Wise innovators select a range of ideas to validate further, with some potentially quick, but valid, wins (bottom-left, easy to do but not the highest impact), along with an essential but limited number of bigger, more demanding ideas (in the top-right, harder to do, but potentially more beneficial).

Whatever filtering process is used, wise innovators will make sure that they are able to:

- Evaluate ideas in a timely way.

- Link evaluations to strategy and, if used, opportunity or challenge briefs.

- Deliver ideas/concepts to validate at a rate that, after likely drop-outs in idea conversion, is aligned to their forecasted pipeline needs (See also Chapters 3 and 5)

- Say 'No' in a constructive way, with explanation of reasons for decisions, or identification of what would need to be addressed before it may become a future 'winner'.

- Say 'Yes' in a way that defines any ongoing hypotheses made for the potential new products/service/process/business and, lists additional assumptions for ongoing validation as the idea is incubated and validated. For example:

    - Value proposition hypothesis:
      - *Customers will buy the new product/service because....*

    - Product/Service assumptions:
      - *We can make the product in...*

    - Channel/distribution assumptions:
      - *Distributors will stock the product because...*

    - Cash flow and funding assumptions:
      - *We can achieve..., with..*

- People and skill assumptions:
  - *We will have no problem attracting the right people.*

- Budget for Validation: = 'x'.

- Timescale for Validation: = 'y'.

Aside from all the above listed good practices for idea selection, most wise innovators will also give a little space to what may initially seem to be really extreme or 'crazy' ideas, that do not necessarily fit with a previously established area of interest. Whilst many of these may end up being judged as strategically irrelevant to pursue, valid opportunities can sometimes arise from off-the-wall ideas and consideration should be given to finding a little space, placing some speculative bets, to allow some of them to develop.

ASSESS YOUR CAPABILITIES

It may now be timely to assess your own organisation's performance with respect to this element. A good way of doing this is to form an assessment team and provide them access to the following 10 questions, which are designed to help you debate and assess your current capabilities and achievements.

How to Assess:

1. Review whether a data-gathering phase will be beneficial, prior to answering the questions, and if so whom to involve.

2. Have each assessment team participant, individually, complete an analysis of each question, circling a score of 1 if they believe that few relevant outcomes/actions can be demonstrated, scoring 5 if they believe you truly have role model practices/outcomes and interpolating between scores 2 to 4 for other conclusions.

3. Undertake a workshop with the aim to create a consensual analysis and score for each question and debate what you think is the next big thing for you to work on.

What to do with the Assessment:

1. Take an average of your agreed scoring of the ten questions and conclude your current 'Reboot Status' for this element. Use this status number to form a base-line for future assessments and to share with interested stakeholders.

2. In areas that you choose to act on, seek insights and good practices that may help you conclude what to do next. We use funds from the sale of this book to fund the Innovation Reboot Project at www.innovationreboot.org. After considering what is the next big thing for you to address, take a look at this website for knowledge, ideas and discussions on each capability element.

Q3.1:
We use an **effective range of creativity tools** and techniques; they work for teams, functions, businesses and external partners.

1. Unable to Demonstrate
2. In Parts
3. Basics Covered
4. Fully Able to Demonstrate
5. We're a Role Model

Q3.2:
We are able to **release the right people**, so they have time to collaborate in ideation activities.

1. Unable to Demonstrate
2. In Parts
3. Basics Covered
4. Fully Able to Demonstrate
5. We're a Role Model

Q3.3:
'Crazy' Ideas are given some space; we do hold unquestionable assumptions about the way we see the world.

1. Unable to Demonstrate
2. In Parts
3. Basics Covered
4. Fully Able to Demonstrate
5. We're a Role Model

Q3.4:
We are **effective at managing external collaborations** and engagement. Our partners are productive.

1. Unable to Demonstrate
2. In Parts
3. Basics Covered
4. Fully Able to Demonstrate
5. We're a Role Model

Q3.5:
We are efficient, timely and effective in **selecting the ideas we want to progress** to validation.

1. Unable to Demonstrate
2. In Parts
3. Basics Covered
4. Fully Able to Demonstrate
5. We're a Role Model

Supporting Evidence:

Supporting Evidence:

Supporting Evidence:

Supporting Evidence:

Supporting Evidence:

Q3.6:
When we select ideas, we produce **clear briefs to guide the validation of the concepts** we choose to explore further.

1. Unable to Demonstrate
2. In Parts
3. Basics Covered
4. Fully Able to Demonstrate
5. We're a Role Model

Q3.7:
When we move ideas to validation **we provide needed resources, in appropriate time frames.**

1. Unable to Demonstrate
2. In Parts
3. Basics Covered
4. Fully Able to Demonstrate
5. We're a Role Model

Q3.8:
When we **say 'No', we do so informatively** and systematically archive the ideas we do not pursue.

1. Unable to Demonstrate
2. In Parts
3. Basics Covered
4. Fully Able to Demonstrate
5. We're a Role Model

Q3.9:
Both our **staff and our external partners think we have great ideation** and engagement processes.

1. Unable to Demonstrate
2. In Parts
3. Basics Covered
4. Fully Able to Demonstrate
5. We're a Role Model

Q3.10:
We obtain potentially valid **ideas in appropriate quantities and time-frames as needed to support our Innovation strategy.**

1. Unable to Demonstrate
2. In Parts
3. Basics Covered
4. Fully Able to Demonstrate
5. We're a Role Model

Supporting Evidence:

Supporting Evidence:

Supporting Evidence:

Supporting Evidence:

Supporting Evidence:

CHAPTER 8

⏻ ELEMENT 4: VALIDATE KEY IDEAS and MAKE BUSINESS PLANS

When an idea is gaining traction and 'growing up', the transitions to 'adult' life always mean accepting different rules and boundaries. Big questions arise, such as; if it works when could the new product/service be launched? What needs to be done to give us confidence that it is right and timely to do this? Will it cannibalise any existing product revenues? A validation programme, with appropriate time frames and resources, needs to be created in order to develop, incubate, 'prove' the idea and answer such questions.

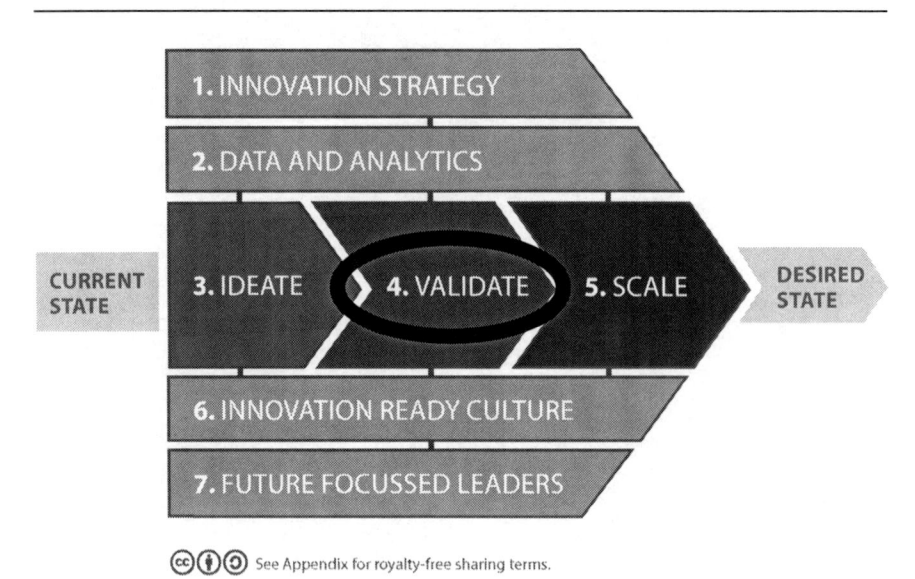

Figure 8.1 Innovation Capability Framework: Validate and Plan

1. INNOVATION STRATEGY
2. DATA AND ANALYTICS
CURRENT STATE — 3. IDEATE — **4. VALIDATE** — 5. SCALE — DESIRED STATE
6. INNOVATION READY CULTURE
7. FUTURE FOCUSSED LEADERS

See Appendix for royalty-free sharing terms.

Validation is a critical element of any wise Innovation ecosystem. Many ideas and concepts will and should fail here. Deciding how to incubate and validate is a difficult decision; too much incubation and validation results in delay and gives competitors opportunities to align, whilst taking an idea directly to full-scale implementation, without some validation, may mean that it is not appropriately sound or valid at its launch.

KEY THINGS WISE INNOVATORS DO: **VALIDATE**

Wise innovators balance validation activity and time, in order to ensure that:

- Element 4.1: **Phased, rapid, validation of the idea takes place** with fast killing of failing ideas and rapid adaption/ progression of ideas and concepts that appear to be of value.

- Element 4.2: **Efficient and timely business planning occurs** and allows the launch of valid ideas and concepts to be agreed and funded as early as possible.

Element 4.1:
PHASED, RAPID, VALIDATION

With the time frames and budgets allocated for a validation phase, wise innovators will aim to iteratively cycle through as many research or testing 'loops' as possible. The aim is to spend a little, learn a lot, reframe and repeat. It's not just about one big experiment/debate/test.

The following diagram shows x5 phase, conceptual, validation cycles (across the top row) and the emphasis that is typically given, by wise innovators, in each phase/cycle.

Figure 8.2 Validation Occurs in Phases

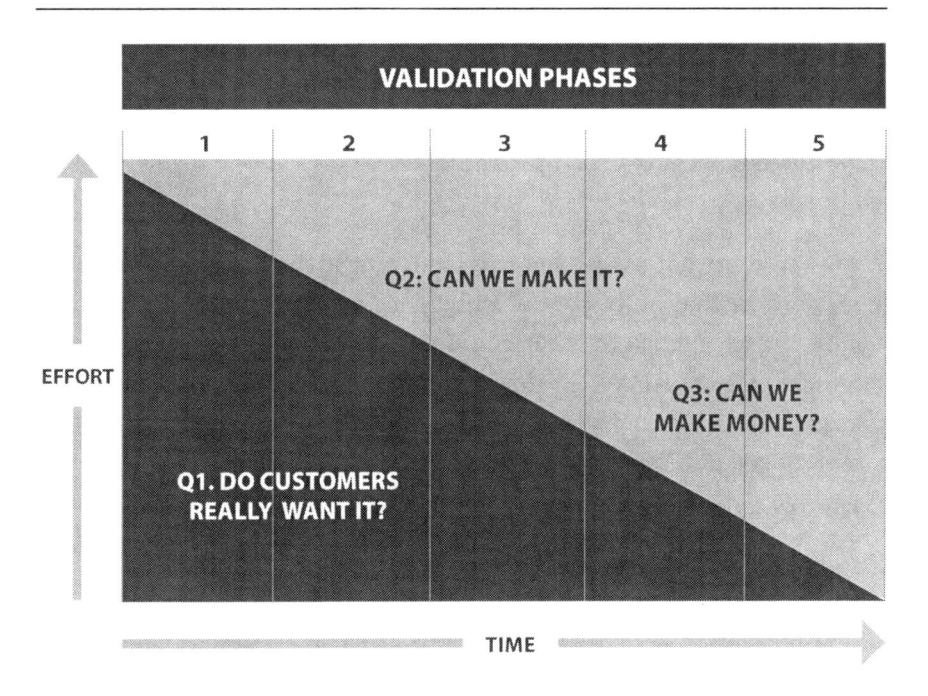

VALIDATION PHASES

| 1 | 2 | 3 | 4 | 5 |

EFFORT

Q1. DO CUSTOMERS REALLY WANT IT?

Q2: CAN WE MAKE IT?

Q3: CAN WE MAKE MONEY?

TIME

In early validation phases the first priority should given to getting early customer validation that the proposed Innovation is really wanted and understanding if it is possible to differentiate the proposed new offering from those of competitors.

In the middle phases emphasis is added to ensuring that it can be made/delivered, with validation input from targeted experts, employees, partners, lawyers and such like.

The final validation cycles need to confirm that a viable business model exists; in these late validation phases wise innovators devote most of their time to confirming that money can be made, by making estimates of market sizes, timings, revenues, costs and profit models.

Within each phase, differing levels of physical validation may be occurring. For example, in Phase 1 it maybe that no physical product or service exists. A validation team may have simply created a phantom website or advertising brochure that enables them to test the concept on a pathfinder customer. As time progresses, increasing emphasis will be put on the physical product or service. So, for example, by Phase 3, an early prototype may have been produced to test production capability, as well as getting ongoing and broader customer feedback. As many 'experiments' as possible, within the timescales and budgets, will be taking place.

Most wise innovators work towards launching at the earliest opportunity. It may well be that, at the time of launch, the product is not fully featured, but by an early launch they can minimise validation costs and maximise potential early revenues. In doing so, they also create opportunity for deeper, fuller, early engagement and feedback from customers and other stakeholders, as a real product/service enters the market. However, this is a never a 'clear-cut' or easy decision; risks exist.

To avoid destroying brand reputations and invalidating Innovation efforts, wise innovators need to convince themselves that they have a minimum viable product/service that is capable of a sound launch. The product/service may not, yet, be fully featured and mature. Some of its hypotheses maybe only partially proven, but wise innovators will have convinced themselves that all foreseeable catastrophic issues, faults and errors have been resolved and that a product of real value exists. If they launch without this confidence they know that the potential loss to their credibility will outweigh the possible gains from an accelerated launch.

A good friend and collaborator, Peter Merrill, in his book 'Innovation Generation'[*1], introduces the metaphor of 'tight-loose' to describe the changing nature of management in differing stages of an innovation pipeline; ideation activities needing a more 'loose' approach to allow creativity, but validation, scaling and launch needing increasingly 'tight' and rigorous business management, in order to ensure cost and profit accountability. In Validation phases, wise innovators will now be applying much more rigour, through structured portfolio and decision management processes, to ensure they are selecting the right projects to progress through this element.

They will use techniques such as Go/Kill scorecards to oversee progress of ideas, as they progress through validation phases. Common Go/Kill scorecard dimensions will include measures of:

- Strategic Fit
- Competitive Advantage
- Market Attractiveness
- Core Competence Leverage
- Feasibility
- Financial reward and risk ratios

[*1] Merrill, P. 'Innovation Generation: Creating an Innovation Process and an Innovative Culture'. American Society for Quality, 2008.

Element 4.2:
EFFICIENT AND TIMELY BUSINESS PLANNING

If all the responses from the validate/test programme (Element 4.1) look good, then business plans need to be made and launch resources assigned (Figure 8.3 –C).

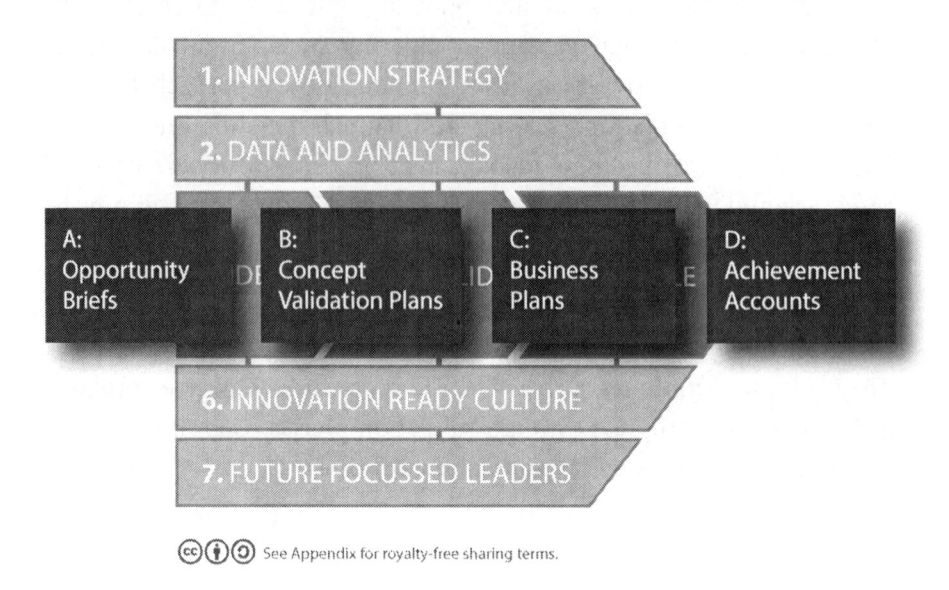

Figure 8.3 Create Simple but Sound Business Plans

1. INNOVATION STRATEGY

2. DATA AND ANALYTICS

| A: Opportunity Briefs | B: Concept Validation Plans | C: Business Plans | D: Achievement Accounts |

6. INNOVATION READY CULTURE

7. FUTURE FOCUSSED LEADERS

See Appendix for royalty-free sharing terms.

Most executive teams, entrepreneurs, innovators and venture capital providers dread receiving (or creating) over-extensive, elaborate and bureaucratic business plans. However, everyone knows that some documented planning is going to be needed to demonstrate viability for funding.

Very early in validation planning, wise innovators will have obtained an outline agreement from their executive team and/or their foreseen capital provider(s), on what data will be needed to make final go/no-go decisions and fund a launch. They will then create the validation program and their business plans in alignment with the funding providers known expectations.

Most wise innovators (and capital providers) consider that a sound business/product launch plan should summarise the validation activities and communicate confidence in three areas:

Figure 8.4 A Sound Business Plan Should Pass Three Tests

[Venn diagram with three overlapping circles labeled:
- Market Need: 1. CUSTOMERS WANT IT
- Economic Value: 3. WE CAN MAKE MONEY
- Operational Solution: 2. WE CAN MAKE IT
Center overlap: VALIDATED OPPORTUNITIES €/£/$/CNY]

Test 1: Customers want it

Wise innovators are able to show that market testing and research indicates that the customers will buy the product/service, that a space amongst competitors exists and that they will likely buy at sensible price points.

Test 2: We can make and deliver it

Wise innovators are able to show that their potential Innovation can be made and delivered in a way that has been shown to provide the foreseen benefits and features and can be consistently produced at the budgeted costs, in planned volumes and in desired time-frames. This will include demonstrating that they can comply with all relevant social, legal and environmental legislation/norms. They will also demonstrate that needed capital, capabilities, resources and people can be obtained and that foreseeable risks, including any IP issues, are understood and can be responded to.

If a new product/service is to be launched in a totally new or spin-off business, then the formation, modus operandi and governance of the new entity will need to be considered and shown to be viable.

If the product/service is being launched within an existing business/operation, or with partners, then any foreseen cultural or integration issues will be highlighted and acted on.

Wise innovators launching new products/services in a larger organisation know that it is important to get early clarity around accounting practices, performance evaluation metrics, overhead allocation, transfer pricing models, brand alignment and such like.

Test 3: We can make money from it

Wise innovators will be able to show that the product/service can secure scalable revenues over time, ('x' years down the 'road' as well as tomorrow) and that profit/value capture is real and at worthwhile levels.

Many innovators will build their revenue forecasting around adaptations of the market 'diffusion' concepts of respected experts such as Rogers*[2], where norms for new Innovation take-up are modelled across differing market maturity phases.

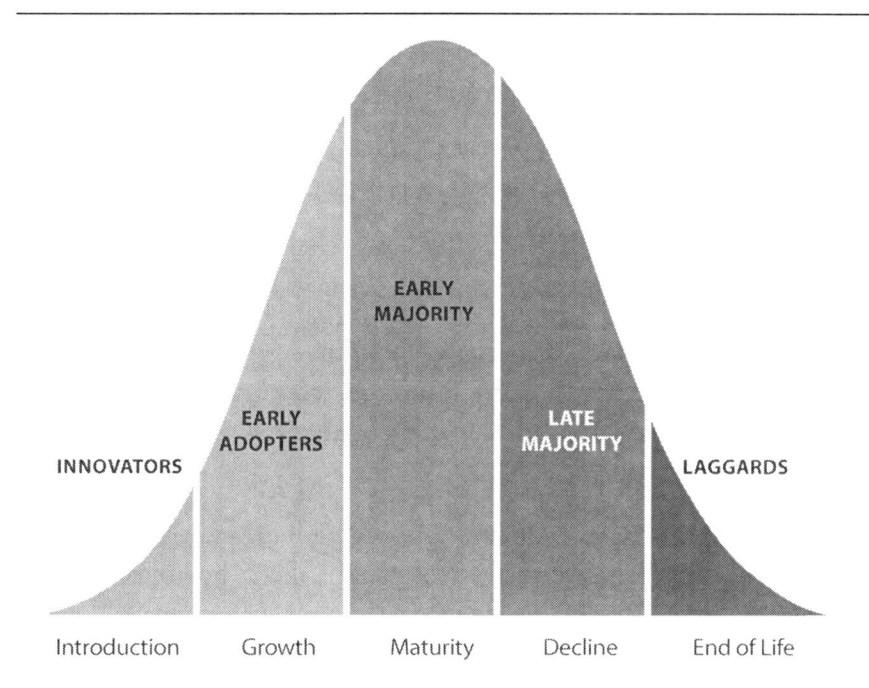

Figure 8.5 Innovation Diffusion and Take-Up over Time

[*2]Rogers, E. (1995). *'Diffusion of Innovations. (4th edn)'*. New York, NY: The Free Press.

Rogers[*2] models the amount of revenue to be obtained ('y' axis) around a normal distribution and a range of purchaser behaviours/categories, but notes that the segmented, category, band widths and timings ('x' axis) will vary according to the nature of the market into which the innovation is being launched.

Commonly, forecast percentages of the amount of total market revenue that each segment/category will deliver are:

Segment 1 Introduction: Introduction and take up by purchasing 'Innovators' in the market is typically forecast at around 2.5% (often over estimated) of the total foreseen market.

Segment 2 Growth: 'Early adopters' represent a segment that is often estimated to represent to the order of 13.5% of the overall market.

Segment 3 Maturity: As the product matures an 'early majority' will likely amount to around 34% of the total market sales. Their confidence being built by feedback from 'early adopters'.

Segment 4 Decline: In later maturity and decline a market will likely have a 'late majority' of 34%.

Segment 5 End of Life: Towards end of life, but not to be forgotten there are 'laggards', typically 16%.

When using such a diffusion/maturity model wise innovators will:

- **Forecast the total life cycle of the Innovation** and the timings of each of the five phases of Innovation adoption (x-axis), i.e. the timing of 'Introduction', 'Growth', 'Maturity', 'Decline' and 'End of Life' phases.

- **Predict the size they believe the overall market** will likely grow to (y-axis). These estimates need to be realistic, however having data driven decision confidence on a totally new concept/Innovation is probably impossible (although some indicators should be projectable based on early feedback in the validation phase. For example, the number of customers and the revenue that will be generated per customer).

- **Hypothesise what will drive the transitions** between each phase and articulate this in their business plans e.g. going from 'early adopters' to 'early majority', which is often a key transition, will be achieved by ...

- **Calculate predicted revenue in each phase.** Typically, by predicting the percentage of market share they foresee they will get in each phase, multiplied by Roger's view of the size of each maturing phase, e.g. 34 per cent for early majority phase, multiplied by their estimate of the overall market size for the Innovation over all phases.

- **Conclude sales and marketing tactics,** showing how customers will find the products/services and what sales and distribution channels will be used in each phase.

ASSESS YOUR CAPABILITIES

It may now be timely to assess your own organisation's performance with respect to this element. A good way of doing this is to form an assessment team and provide them access to the following 10 questions, which are designed to help you debate and assess your current capabilities and achievements.

How to Assess:

1. Review whether a data-gathering phase will be beneficial, prior to answering the questions, and if so whom to involve.

2. Have each assessment team participant, individually, complete an analysis of each question, circling a score of 1 if they believe that few relevant outcomes/actions can be demonstrated, scoring 5 if they believe you truly have role model practices/outcomes and interpolating between scores 2 to 4 for other conclusions.

3. Undertake a workshop with the aim to create a consensual analysis and score for each question and debate what you think is the next big thing for you to work on.

What to do with the Assessment:

1. Take an average of your agreed scoring of the ten questions and conclude your current 'Reboot Status' for this element. Use this status number to form a base-line for future assessments and to share with interested stakeholders.

2. In areas that you choose to act on, seek insights and good practices that may help you conclude what to do next. We use funds from the sale of this book to fund the Innovation Reboot Project at www.innovationreboot.org. After considering what is the next big thing for you to address, take a look at this website for knowledge, ideas and discussions on each capability element.

Q4.1:
All validation stage projects have **appropriate accountability, sponsorship and support from executive managers.**

1. Unable to Demonstrate
2. In Parts
3. Basics Covered
4. Fully Able to Demonstrate
5. We're a Role Model

Q4.2:
We always test **customer, operational and economic assumptions in our validation programmes.**

1. Unable to Demonstrate
2. In Parts
3. Basics Covered
4. Fully Able to Demonstrate
5. We're a Role Model

Q4.3:
We are good at **stopping applying resources to failing projects,** prepared to acknowledge failures and **make 'fast kills'.**

1. Unable to Demonstrate
2. In Parts
3. Basics Covered
4. Fully Able to Demonstrate
5. We're a Role Model

Q4.4:
If an Innovation **might cannibalise existing revenues**, we still take a mature look at the overall possible benefits.

1. Unable to Demonstrate
2. In Parts
3. Basics Covered
4. Fully Able to Demonstrate
5. We're a Role Model

Q4.5:
Our strategy guides **when to scale in our existing business** and when to use a new organisation, spin-off or joint venture.

1. Unable to Demonstrate
2. In Parts
3. Basics Covered
4. Fully Able to Demonstrate
5. We're a Role Model

Supporting Evidence:

Supporting Evidence:

Supporting Evidence:

Supporting Evidence:

Supporting Evidence:

Q4.6:
Our validation pipeline progresses at **strategically appropriate speeds.** We are rarely disrupted by others moving faster.

1	Unable to Demonstrate
2	In Parts
3	Basics Covered
4	Fully Able to Demonstrate
5	We're a Role Model

Q4.7:
When making launch plans we **evaluate likely alternative offerings of our competitors** and forecast accordingly.

1	Unable to Demonstrate
2	In Parts
3	Basics Covered
4	Fully Able to Demonstrate
5	We're a Role Model

Q4.8:
We always hold **post validation project learning** reviews.

1	Unable to Demonstrate
2	In Parts
3	Basics Covered
4	Fully Able to Demonstrate
5	We're a Role Model

Q4.9:
Our validation pipeline is filled with a **balanced portfolio of projects** that align with our innovation strategy.

1	Unable to Demonstrate
2	In Parts
3	Basics Covered
4	Fully Able to Demonstrate
5	We're a Role Model

Q4.10:
All involved stakeholders think our **new business planning and funding process is effective and efficient.**

1	Unable to Demonstrate
2	In Parts
3	Basics Covered
4	Fully Able to Demonstrate
5	We're a Role Model

Supporting Evidence:

Supporting Evidence:

Supporting Evidence:

Supporting Evidence:

Supporting Evidence:

CHAPTER 9

⏻ ELEMENT 5: SCALE THE BEST NEW IDEAS OR BUSINESSES

An organisation or entrepreneur may have come a long way, discovered some exciting unmet needs, ideated well, validated some potentially world-beating ideas and convinced themselves a viable business model exists, but all such efforts can be destroyed by lack of attention to the effective scaling of a new idea into its market place.

No Innovation is of any value until it's been successfully deployed. This capability framework element is about reviewing the methods used to launch and scale a new 'business'. Wise innovators know that scaling up fast and flexibly is often a determinant of subsequent successful growth. Many who do not recognise this fail and sit, frustrated, on the side-lines to watch others 'fast-follow' and exploit their ideas.

Figure 9.1 Innovation Capability Framework: Scaling

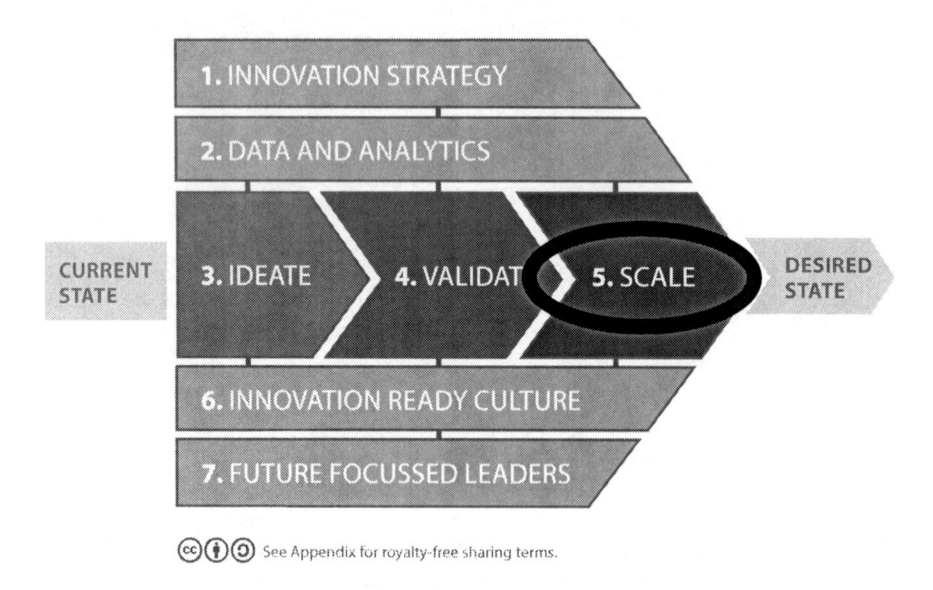

See Appendix for royalty-free sharing terms.

Scaling effectively requires speed and agility to be derived from, what are likely to be, new teams of people working in new, or significantly revised, operational processes.

KEY THINGS WISE INNOVATORS DO: **SCALE**

When new product/services are launched wise innovators will ensure that they are:

- Element 5.1: **Not constrained by too much 'process glue'.**
  They will build, around the new product/service/business, the minimum necessary level of controls and processes, with the right amount of 'process glue' needed to ensure assurance of

operations, but do so in a way that does not constrain their ability to react and change.

⏻ Element 5.2: **Obtaining the right people** or partners that they need to build motivated teams to manage the new Innovations and operations.

⏻ Element 5.3: **Hyper agile at launch.** They will carefully monitor early outcomes and iterate or 'pivot' the new business direction, in a timely way, when needed.

Element 5.1:
PROCESSES – How much glue?

Formal and informal management systems will of necessity be created as a new business or product is launched. Wise innovators know that finding a balance between the level of process/system, that delivers consistency and assurance of operation, but with a needed agility and freedom to react, is something that requires management time and attention and is easily overlooked during the pressures of early growth. They know that if they do not take time to review this well, that the long-term consequences, to their efficiency or effectiveness of their new operation can be significant.

Trigger points, such as the number of employees/contractors engaged in a new venture, can act as useful milestones to indicate a need to stand back and review progress with management systems.

Figure 9.2 shows examples of such review points. The 'critical-mass' described in the diagram and exampled by number of employees, may vary according to the nature of the product/organisation.

Figure 9.2 Critical Mass Trigger Points in Organisational Growth

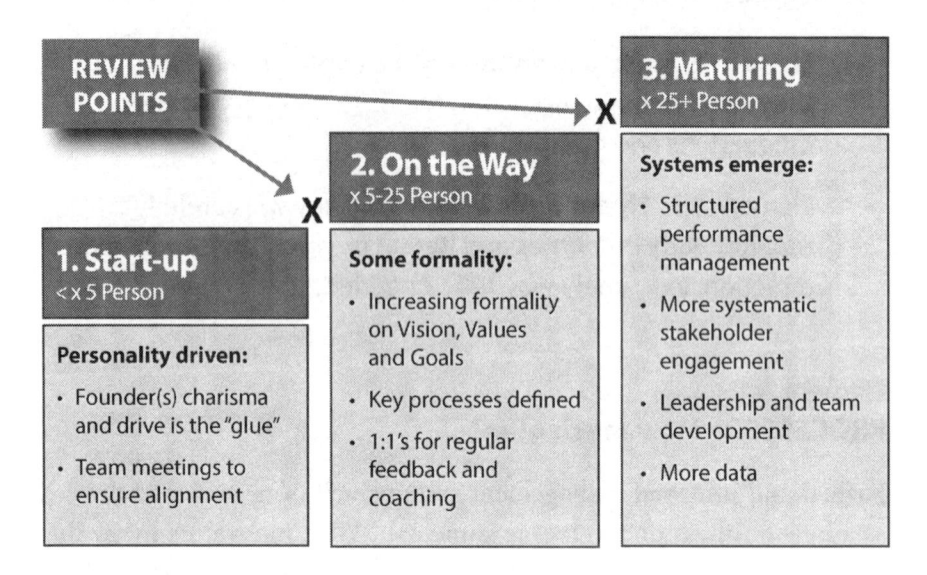

REVIEW POINTS

1. Start-up
< x 5 Person

Personality driven:
- Founder(s) charisma and drive is the "glue"
- Team meetings to ensure alignment

2. On the Way
x 5-25 Person

Some formality:
- Increasing formality on Vision, Values and Goals
- Key processes defined
- 1:1's for regular feedback and coaching

3. Maturing
x 25+ Person

Systems emerge:
- Structured performance management
- More systematic stakeholder engagement
- Leadership and team development
- More data

Wise innovators take time to review and adapt needed systems as they scale in organisational size. Common risk-review points are shown in Figure 9.2 and can be summarised as the points of transition between:

Start-Up:

Scaling from 1 to 5 employees can often be achieved with no more than the management 'charisma' of the founder manager(s); little additional process 'glue' will be needed. Simply setting operational standards, a few key 'dashboard' measures (see Elements 5.3 and 7.3) and reviewing – with key team members – a weekly list of key goals to accomplish for the following period, may be enough to communicate what is expected and ensure that everyone knows how they will work together to realise it.

On-the-Way:

The transition from 5–25 engaged employees is often perceived as a risk area. The enlarging team now needs additional coordination, but the founder(s) are busy growing the business and cannot always find the time to provide it. Serial entrepreneurs often say that, in hindsight they wished they had devoted more management time to overseeing this 'on-the-way' phase, as it is here that both good and bad, culturally definitive, precedents can be set and their ability for future rapid growth becomes defined.

Maturing:

When a team size gets to 25+ employees, the need for system 'glue' typically ramps up to a higher level. Structured governance and performance management methods will be demanded by an increasing number of customers and other stakeholders, who will now be engaged. For entrepreneur mangers, this is also about creating legacy systems for the business, as without them those seeking to sell their new businesses may find that prospective purchases consider the business to be too dependent on a small number of individuals. Wise innovators will be paying particular attention to increased formalisation of operational methods and more systematic and data driven ways of understanding the needs of their new stakeholders (key customers, capital providers, partners, etc.) so as to help understand and manage their needs and expectations.

Element 5.2:
GETTING THE RIGHT PEOPLE

Wise innovators place great importance on getting good people, (and/or hiring the right freelancers or partners). They recognise that the new business operation, or team, has the great 'gift' of being able to build the 'culture' it needs, and embed desired values and norms, right from the start.

With respect to recruiting aims, at all levels, *'hire for tomorrow not just today'*, is an often-repeated mantra. Wise innovators will have developed their recruiting 'pitch', and will have been working on fulfilling their needs for many months in advance of the launch. They will place great emphasis on attracting those who they believe can fully align to their aims, values and new ways of working. To illustrate this point, a very well-known, innovative, global technology giant asks itself the following, four weighted questions when assessing potential recruits.

Q1: *'Are they agile thinkers?'* (40 per cent weighting)
Q2: *'Are they self-managed and aligned to our values?'* (30 per cent),
Q3: *'Are they technically competent?'* (20 per cent),
Q4: *'Can they think big and crazy?'* (10 per cent).

Noteworthy here is that technical competency evaluation (Q3) is third, not first on the list. Its omission can often be fixed by training or induction solutions, whereas, resolving omissions in areas such as agile thinking (Q1) and self-management/values (Q2) are not such easy fixes.

Another similar perspective can be seen in CEO views at Apple:

'You look for people that are not political. People that are not bureaucrats. People that can privately celebrate the achievement, but not care if their name is the one in the lights. There are greater reasons to do things. You look for wicked smart people. You look for people who appreciate different points of view. People who care enough that they have an idea at 11 at night and they want to call and talk to you about it. Because they're so excited about it, they want to push the idea further.' –Tim Cook, CEO, Apple Inc., speaking at Duke Fuqua Business School on the 25th anniversary of his MBA in 2013.

As growth occurs, maintaining such worthy principles can be difficult to uphold, particularly if new customers are knocking on your door and you do not have capability to produce their product/service – but it is often cited by wise innovators as being key to long-term success.

With regard to recruiting managers, many wise innovators comment that there is no substitute for using people who have scaled a business before and will ensure that at least one executive, with such skills, is engaged. If they are wise they will also challenge the 'old-hand' to work with some less experienced, but high potential managers, who may then become part of a resource pool for future launches. Sometimes the founders or innovators from the earlier phases are not going to be the best people to manage the scaled-up operations. This can create tensions if expectations are not managed carefully.

Figure 9.3 Managing Team Expectations

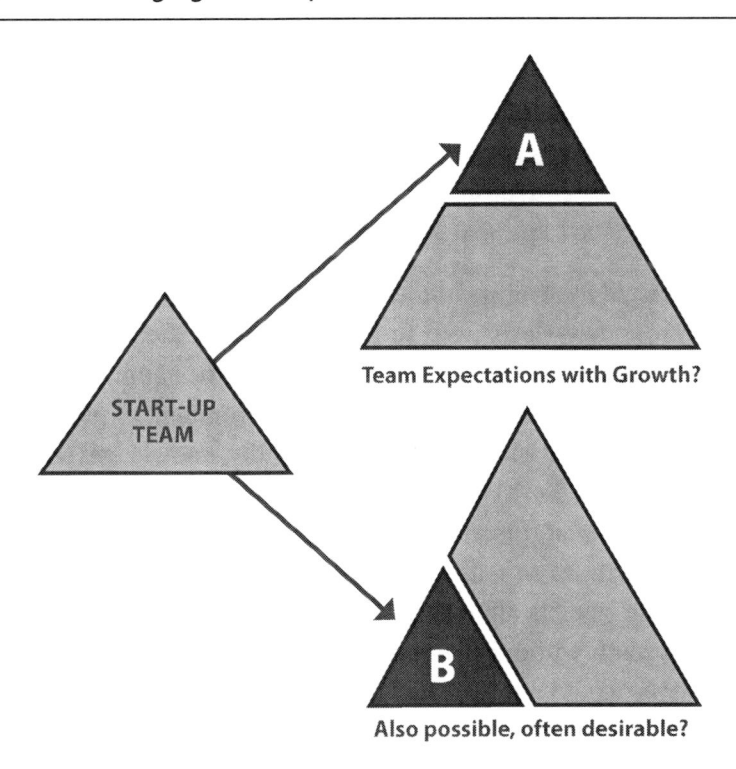

Team Expectations with Growth?

Also possible, often desirable?

The team that led the validation phase may well believe that they should be following path 'A' to the top of the 'pyramid', the 'boardroom' of the launched business. But for both them and the new enterprise, option 'B' is also often valid, as is the more dramatic option of them moving on from the business. Wise innovators know that serial entrepreneurs, or in-company innovators are not always the best managers to take new-start businesses to maturity and they often become frustrated when they try to do so. Wise innovators openly review the merits of all options and ensure that both the expectations and competencies of key players are not misplaced.

Maintaining a balanced and diverse team is a further risk often cited by new-start businesses. There is a danger that the founder management team of a new-start entity, or a spin-off company, may recruit newcomers in a way that attempts to clone others into their own 'mould', i.e. they do not maximise the benefits that, for example, gender, age and cultural balance can bring.

Reid Hoffman (founder of LinkedIn) was recently quoted as saying *'Don't be ageist; broad or deep experiences can add great value, but ask yourself if what you are getting (recruiting) is 20 years' of experience, or is it actually one year of experience repeated 20 times'.*

Most experienced innovators intuitively believe that the 'grunt' and 24/7 focus that most new-starts need to have applied to their businesses, is probably better given by those in their younger years. Families often take them over in midlife, but there are increasing numbers of post-midlife opportunities for those with the adrenaline, agility, knowledge (assuming that they have kept it fresh) and motivation to return to frontline growth. Hoffman's point is that this should not be ignored, and those who return in such roles often do so with the wisdom of past errors incorporated. There is no one size fits all answer. Facebook would not have achieved their rapid growth without the youthful Mark Zuckerburg, but equally Apple probably would not have had their success in the last two decades without a midcareer Steve Jobs, having come back for what was, regrettably, too short a last term.

Wise innovators know that neither age or youth, gender or race' limits a person's ability to contribute to a new businesses growth; they recognise that diversity can help foster Innovation. They also know that if personal motivation (in the potential recruit) is lacking and if the business's and their personal, aims and values are not shared, then issues will arise. They seek to recruit a motivated, agile, value-driven and diverse workforce with these concerns in mind.

Element 5.3:
HYPER-AGILITY AT LAUNCH

Wise innovators know they will not get everything right; launches are never perfect. A much needed skill is an ability to be able to sense any missed opportunity, identify what is not going as expected and address the underlying issues in a timely way.

To sense such risk/opportunity, most innovators will put in place a 'launch dashboard' of key business measures to cover areas such as:

- **Customer Engagement, Acquisition and Retention:**
  Are volumes of leads/web site use/social media responses and other sales or marketing indicators to plan? Are responses timely? Is customer acquisition to plan?

- **Customer Insight Indicators:**
  Why are our customers buying or expressing interest, why are others not?

- **Channel Insight Indicators:**
  Is channel effectiveness and the costs of customer acquisition as budgeted?

- **Competitor Insight Indicators:**
  How are competitors reacting?

- **Cash Flow Metrics:**
  Are funding milestones met? Do cost reductions or other monetisation opportunities exist?

- **Key Process Statistics:**
  Are production and delivery standards being met? Are partners performing as expected?

Initially, data may be limited, but even so, wise innovators will always regularly and systematically assess early achievements. Such reviews may lead them to conclude one of three outcomes:

1. That they got it all totally right (nice, but not common).

2. That they have totally failed (not nice, but thankfully, also not common at this late stage, if the validation work was robust).

3. That they need to react and change their plans (most likely).

Most launches involve a need to 'iterate' (make minor corrections or 'morph' the current plan/model/market). For example, Google developed three social networking platforms before arriving at Google+.

Sometimes early market feedback will indicate a need for dramatic redirection of thinking. The product/service may be seen to have value, but not in the market or way in which it was originally foreseen. Many launches end up with a redirecting of a launched product/business, with new aims or markets in mind. Many examples can be listed for well-known Innovations that have successfully 'pivoted' at launch, for example, YouTube was originally a video dating site, and Twitter started as a website (Odeo) for sharing of podcasts.

Wise innovators know that agility, particularly immediately post–launch, is very important. Any decisions to 'iterate' or 'pivot' need to be taken and acted on very quickly, as at this point, the new product/service is out/visible in the market place and potential competitors maybe 'fast following'. Those innovators who are agile and iterate or pivot quickest, are often those who succeed the most.

ASSESS YOUR CAPABILITIES

It may now be timely to assess your own organisation's performance with respect to this element. A good way of doing this is to form an assessment team and provide them access to the following 10 questions, which are designed to help you debate and assess your current capabilities and achievements.

How to Assess:

1. Review whether a data-gathering phase will be beneficial, prior to answering the questions, and if so whom to involve.

2. Have each assessment team participant, individually, complete an analysis of each question, circling a score of 1 if they believe that few relevant outcomes/actions can be demonstrated, scoring 5 if they believe you truly have role model practices/outcomes and interpolating between scores 2 to 4 for other conclusions.

3. Undertake a workshop with the aim to create a consensual analysis and score for each question and debate what you think is the next big thing for you to work on.

What to do with the Assessment:

1. Take an average of your agreed scoring of the ten questions and conclude your current 'Reboot Status' for this element. Use this status number to form a base-line for future assessments and to share with interested stakeholders.

2. In areas that you choose to act on, seek insights and good practices that may help you conclude what to do next. We use funds from the sale of this book to fund the Innovation Reboot Project at www.innovationreboot.org. After considering what is the next big thing for you to address, take a look at this website for knowledge, ideas and discussions on each capability element.

Q5.1:
We create **just enough structure and process** in our launch operations.

1. Unable to Demonstrate
2. In Parts
3. Basics Covered
4. Fully Able to Demonstrate
5. We're a Role Model

Q5.2:
At the needed time, **we obtain the 'right' people** to enable us to launch and grow our new business/products.

1. Unable to Demonstrate
2. In Parts
3. Basics Covered
4. Fully Able to Demonstrate
5. We're a Role Model

Q5.3:
When recruiting teams for our new operations, we do so with appropriate **gender, age and cultural balance.**

1. Unable to Demonstrate
2. In Parts
3. Basics Covered
4. Fully Able to Demonstrate
5. We're a Role Model

Q5.4:
We make time to provide **feedback, coach and develop people in their new teams.**

1. Unable to Demonstrate
2. In Parts
3. Basics Covered
4. Fully Able to Demonstrate
5. We're a Role Model

Q5.5:
We track a **sensible set of key metrics to oversee the reality of what the new product/service is achieving.**

1. Unable to Demonstrate
2. In Parts
3. Basics Covered
4. Fully Able to Demonstrate
5. We're a Role Model

Supporting Evidence:

Supporting Evidence:

Supporting Evidence:

Supporting Evidence:

Supporting Evidence:

Q5.6:
We are able to get **good insights into why customers are buying or expressing interest,** and why others are not.

1. Unable to Demonstrate
2. In Parts
3. Basics Covered
4. Fully Able to Demonstrate
5. We're a Role Model

Q5.7:
We achieve an **acceptable control of cash flow;** any needed ongoing funding is secured.

1. Unable to Demonstrate
2. In Parts
3. Basics Covered
4. Fully Able to Demonstrate
5. We're a Role Model

Q5.8:
We **keep track of what competitors are doing** when we are scaling.

1. Unable to Demonstrate
2. In Parts
3. Basics Covered
4. Fully Able to Demonstrate
5. We're a Role Model

Q5.9:
Our launch **review processes are agile, timely and effective.** We iterate, redirect or 'pivot' when needed.

1. Unable to Demonstrate
2. In Parts
3. Basics Covered
4. Fully Able to Demonstrate
5. We're a Role Model

Q5.10:
Our past **launches have been successful and on time.**

1. Unable to Demonstrate
2. In Parts
3. Basics Covered
4. Fully Able to Demonstrate
5. We're a Role Model

Supporting Evidence:

Supporting Evidence:

Supporting Evidence:

Supporting Evidence:

Supporting Evidence:

CHAPTER 10

⏻ ELEMENT 6: INNOVATION READY CULTURE

People embedded in a strong organisational culture hold widely shared, tacit understandings about how they should think and act. Wise innovators invest time to understand such unwritten beliefs and cultural norms; they know that if they are conflicting with the organisation's Innovation aims, then progress will be difficult and a negative underlying culture will 'trump' Innovation strategy every time.

Figure 10.1 Innovation Capability Framework: Innovation Ready Culture

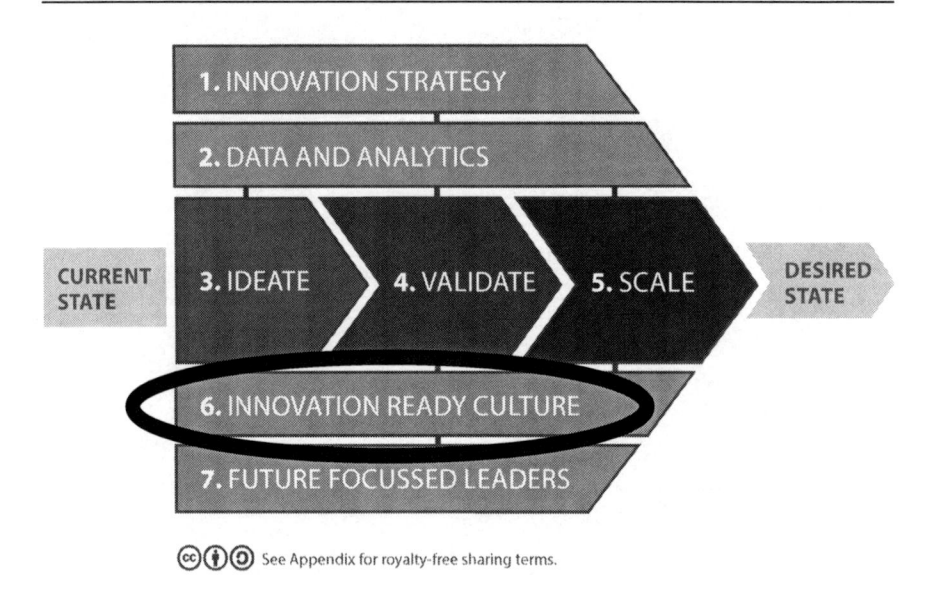

CURRENT STATE

1. INNOVATION STRATEGY
2. DATA AND ANALYTICS
3. IDEATE
4. VALIDATE
5. SCALE
6. INNOVATION READY CULTURE
7. FUTURE FOCUSSED LEADERS

DESIRED STATE

See Appendix for royalty-free sharing terms.

Wise innovators know that if personal, team and organisational goals/systems, are not aligned in a way that allows (or even demands) space for, and participation in Innovation, then progress will be slow.

KEY THINGS WISE INNOVATORS DO:
INNOVATION READY CULTURE

Wise innovators will act on commonly accepted principles of organisational and behavioural change, including:

- Element 6.1: **Making the imperatives clear,** by ensuring that everyone in the organisation knows why they, and the organisation, must innovate and change.

- Element 6.2: **Developing participation competencies and opportunities** by providing awareness, training and innovation 'tools' in order to ensure that everyone has an opportunity to participate in and/or share their views on, Innovation-related activities.

⏻ Element 6.3: **Providing feedback and re-enforcement** to recognise progress and success, confront any non-value adding behaviours and motivate ongoing Innovation enabling activities.

Element 6.1:
MAKING THE IMPERATIVES CLEAR

Although it is difficult to alter deep-seated beliefs and behaviours quickly, wise innovators know that making the imperatives (the need for Innovation and change) clear is a must-do practice and, if reinforced with opportunity to participate and feedback, can help over time to change beliefs and behaviours.

They communicate, with passion and urgency, why Innovation is essential and contextualise what it means for each individual in the organisation.

Figure 10.2 Building a Supportive Culture

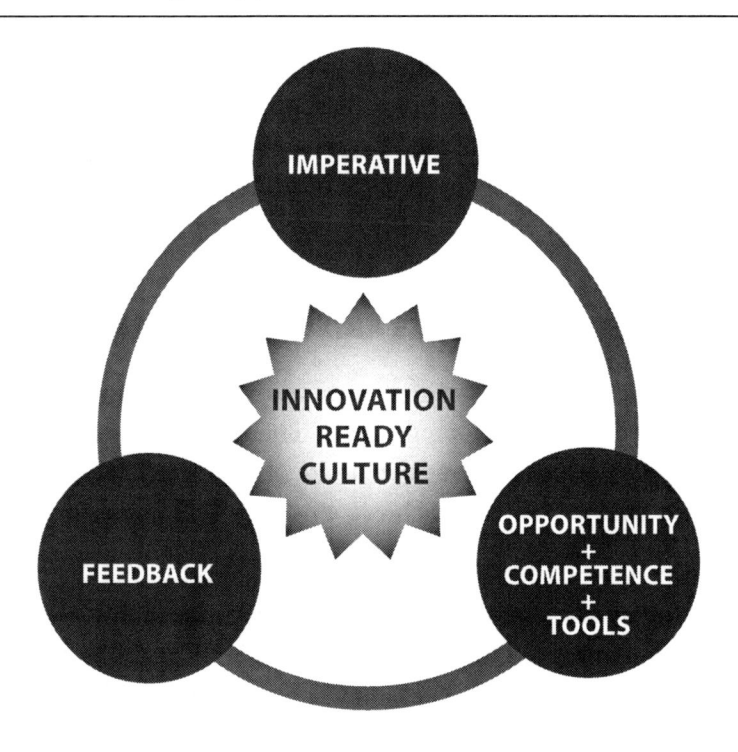

To communicate the imperatives, they will consider practices such as:

- **Communicating, in their own organisational context**, their responses to the opportunities and risks identified in Chapters 1–5 of this book. When doing this they will often have 'message' content that describes imperatives based on fact and reason (*e.g. 'We will not survive past 2020 without new products'*), but also will add focus based on emotion and vision (*e.g. An altruistic need as well as a company need. 'There is a global need to improve medicare in ....., we can eradicate this by ....'*).

- **Making clear a personalised 'What's in it for me'** summary of the opportunities and risks, articulated in a way that has meaning for employees/partners at all levels and indicates what is most critical for them to do.

- **Sharing success stories** of the outcomes of insightful Innovation achievements and failures, both within the organisation and from the external good practice of others, including illustrating the risks of competitor's successes.

- **Using domain/subject experts** to broadcast ongoing awareness information on specific Innovation opportunities/risks – via Internet portals, social media or any other media platform that will broadcast widely in subjects of strategic interest.

Element 6.2:
DEVELOPING COMPETENCIES AND OPPORTUNITIES

Wise innovators recognise that they need to give 'permission' to innovate by providing 'tools', developing competencies to innovate and then creating opportunities for everyone to be able to participate and share their ideas and/or views. To achieve this they will consider practices such as:

- **Doing needs assessments** by assigning responsibilities to key senior manager(s) to undertake Innovation Capability and related training needs assessments. For example by initially undertaking reviews as described in Chapter 4, so that Innovation capability improvement plans can be put in place in all parts of the organisation.

- **Provision of appropriate training** and Innovation 'tools' that can be used to build a common, organisation wide, Innovation 'language' and collaboration methodology.

- **Assigning Innovation mentors** to encourage engagement and help people to take the first steps.

- **Continually update all employees on Innovation projects** in development so that anyone can comment on them.

- **Ensuring Leaders 'show not sell'** by ensuring that leaders and influencers are role modelling desired participation in Innovation processes/forums and they use any Innovation tools/systems that may exist.

- **Having focussed engagement campaigns**, for example at lower organisational levels, having campaigns for ideas around a specific 'opportunity brief' or if that is not relevant a general theme *(e.g. 'For the next 3 months we want to see Innovation in....').*

Element 6.3:
PROVIDING PERSONAL FEEDBACK TO EVERYONE

Wise innovators expect both employees (peer to peer) and leaders to provide feedback that will reinforce Innovation enabling behaviours and/or confront any non value-adding actions. To achieve this they will consider practices such as:

- **Making praise a habit** by using formal and informal reinforcements to 'recognise' (for example with thank you's and small celebratory awards) and/or 'reward' (with payments or gifts of tangible value) the Innovation successes, and efforts, of both individuals and teams.

- **Making Innovation part of employee performance reviews** and linking advancement decisions to demonstrated Innovation competence/success and 'living' the company's Innovation values.

- **Using feedback/coaching to confront, then support,** anyone who is not engaging in the desired way, clarifying to them, the specific implications of their behaviour.

- **Tracking Innovation-enabling measures** for teams and individuals and giving regular feedback on progress (See Chapter 11 for more on Innovation enabling measures).

- **Sensing employee mood** and tolerance for Innovation on a regular basis and acting on issues arising. For large organisations, this does not necessarily mean, yet another question in an annual employee survey. A weekly survey or mood indicator can often add value.

- **Building a tolerance for smart failure**, effective failures should be planned for and accepted. Failing better (i.e. faster, cheaper and with good learning) should be a business aim. Wise innovators develop an environment in which employees can fail on projects, with their egos and careers remaining appropriately intact. For example, at W.L.Gore (GoreTex), managers celebrate, with a party, the killing (stopping) of a project: the goal being to take time to recognise the efforts of the team, give and get feedback and learn.

ASSESS YOUR CAPABILITIES

It may now be timely to assess your own organisation's performance with respect to this element. A good way of doing this is to form an assessment team and provide them access to the following 10 questions, which are designed to help you debate and assess your current capabilities and achievements.

How to Assess:

1. Review whether a data-gathering phase will be beneficial, prior to answering the questions, and if so whom to involve.

2. Have each assessment team participant, individually, complete an analysis of each question, circling a score of 1 if they believe that few relevant outcomes/actions can be demonstrated, scoring 5 if they believe you truly have role model practices/outcomes and interpolating between scores 2 to 4 for other conclusions.

3. Undertake a workshop with the aim to create a consensual analysis and score for each question and debate what you think is the next big thing for you to work on.

What to do with the Assessment:

1. Take an average of your agreed scoring of the ten questions and conclude your current 'Reboot Status' for this element. Use this status number to form a base-line for future assessments and to share with interested stakeholders.

2. In areas that you choose to act on, seek insights and good practices that may help you conclude what to do next. We use funds from the sale of this book to fund the Innovation Reboot Project at www.innovationreboot.org. After considering what is the next big thing for you to address, take a look at this website for knowledge, ideas and discussions on each capability element.

Q6.1:
Our people and our partners **understand the Innovation imperatives** we face and engage with us to address them.

1. Unable to Demonstrate
2. In Parts
3. Basics Covered
4. Fully Able to Demonstrate
5. We're a Role Model

Q6.2:
We use **collaboration and Innovation tools**, effectively, across teams, functions and with external partners.

1. Unable to Demonstrate
2. In Parts
3. Basics Covered
4. Fully Able to Demonstrate
5. We're a Role Model

Q6.3:
Our **leaders coach their employees** and provide feedback to individuals on innovation enabling practices.

1. Unable to Demonstrate
2. In Parts
3. Basics Covered
4. Fully Able to Demonstrate
5. We're a Role Model

Q6.4:
We have evaluated our Innovation **capabilities and put in place improvement plans with training and support** as needed.

1. Unable to Demonstrate
2. In Parts
3. Basics Covered
4. Fully Able to Demonstrate
5. We're a Role Model

Q6.5:
Everyone has the **opportunity to participate** in activities related to our Innovation aims.

1. Unable to Demonstrate
2. In Parts
3. Basics Covered
4. Fully Able to Demonstrate
5. We're a Role Model

Chapter 10: Element 6 **INNOVATION READY CULTURE**

Supporting Evidence:

Supporting Evidence:

Supporting Evidence:

Supporting Evidence:

Supporting Evidence:

Q6.6:
People at all levels will **challenge their peers** if they are not living our 'values' or participating as expected.

1. Unable to Demonstrate
2. In Parts
3. Basics Covered
4. Fully Able to Demonstrate
5. We're a Role Model

Q6.7:
Advancement, reward and recognition are appropriately linked to our Innovation aims.

1. Unable to Demonstrate
2. In Parts
3. Basics Covered
4. Fully Able to Demonstrate
5. We're a Role Model

Q6.8:
Leaders and subject experts regularly communicate or blog to both employees and partners, about trends and ideas.

1. Unable to Demonstrate
2. In Parts
3. Basics Covered
4. Fully Able to Demonstrate
5. We're a Role Model

Q6.9:
We support prudent risk taking and are **accepting of smart failures.** We learn from failures in a way that leaves intact the ego's and career progression prospects of those involved.

1. Unable to Demonstrate
2. In Parts
3. Basics Covered
4. Fully Able to Demonstrate
5. We're a Role Model

Q6.10:
We can see the benefits from **collaboration across units, businesses, partners and subsidiaries.**

1. Unable to Demonstrate
2. In Parts
3. Basics Covered
4. Fully Able to Demonstrate
5. We're a Role Model

Chapter 10: Element 6 INNOVATION READY CULTURE

Supporting Evidence:

Supporting Evidence:

Supporting Evidence:

Supporting Evidence:

Supporting Evidence:

CHAPTER 11

⏻ ELEMENT 7: FUTURE FOCUSSED LEADERS

Innovation requires leadership ambidexterity, it demands organisations and managers that are able to maintain the core business, to exploit what they have, but also, simultaneously, to identify new opportunities, collaborate, generate ideas and innovate. Easy to say, but difficult to achieve, when leaders in many organisations are immersed in the fire-fighting of daily crises and where keeping things on track and under control are the key activities they have historically practiced.

Figure 11.1 Innovation Capability Framework: Future Focussed Leaders

```
                    1. INNOVATION STRATEGY
                    2. DATA AND ANALYTICS
CURRENT                                                              DESIRED
STATE       3. IDEATE      4. VALIDATE      5. SCALE                 STATE
                    6. INNOVATION READY CULTURE
                    7. FUTURE FOCUSSED LEADERS
```

See Appendix for royalty-free sharing terms.

Wise innovators realise that to be successful a new leadership mind-set is often needed. They devote time and resources to support leaders to make needed transitions.

KEY THINGS WISE INNOVATORS DO:
FUTURE FOCUSSED LEADERS

Wise innovators support leaders to make transitions and develop new skills, so they are able to:

- Element 7.1: **Demonstrate leadership ambidexterity**, by developing skills that enable them not only able to maintain the core business but also, simultaneously, to be able to free time to devote to managing the future.

- Element 7.2: **Use collaborative leadership styles when needed**, in order to maximise the broad engagement of diverse stakeholders in support of the organisations Innovation aims.

- Element 7.3: **Personally manage and coach Innovation**

enabling activities, with skills to measure, oversee, support and coordinate those whom they are leading to innovate.

Element 7.1:
LEADERSHIP AMBIDEXTERITY

Wise innovators know that their leaders of the future will need to possess skill sets that enable them to make sense, meaning and opportunity out of the flood of signals received from their strategic networks and then subsequently, to be able to lead diverse, cross-functional (in-company) and/or open networks (outside company) to collaborate, create or obtain ideas and enable Innovation. For many leaders this means making significant transitions towards more future focussed leadership behaviours and practices. It means letting go of past practices, where just optimising what they were responsible for managing was often enough, to a future where both optimising and innovating what they do will become part of their daily lives.

Figure 11.2 Transitions in a Future Focussed Leader

WAS HAPPY WITH:	NOW THRIVES ON:
Certainty, and a status quo	Ambiguity, uncertainty and risk
Optimising everything (if we are efficient and compliant we are good)	Optimising what is core, at the same time developing new
Decisions made from experience, intuition and "gut" feel	Seeking broad data and insights to reinforce decisions
Telling others what to do	Collaborating on deciding the right actions
Managing just own team	Developing global teams, virtual teams, partners and networks
Accepting diversity	Valuing and optimising diversity
Never failing (getting a result in any way that works)	Getting results in the right way and accepting that smart failure is ok

Many wise innovators put in place educational and coaching programmes to help leaders develop new skills such as 'questioning', 'observing', 'networking', 'experimenting' and 'associating'. These five skills often being quoted as the key skills of disruptive innovators[1].

[1] Dyer, J., Gregersen, H., and Christensen, C., 'The Innovator's DNA: Mastering the Five Skills of Disruptive Innovators'. Brilliance Corporation, 2012.

Element 7.2:
A COLLABORATIVE STYLE, WHEN NEEDED

The role of most leaders has, or is becoming, extended outside the organisation. Most leaders are no longer in 'control' of all the participants with whom they need to work with. They not only have to lead and manage their own internal teams but also bring their leadership skills to bear to help optimise the relationships with the external networks or partners with whom they collaborate to create ideas or products.

Collaboration is therefore a skill that most wise innovators recognise as needed. To lead effectively, leaders of the future will need to facilitate the achievement of their organisation's needed outcomes, but often will have to do so without an authority to directly dictate or demand what participants do. Wise innovators invest time to support leaders to do this effectively, but recognise that collaborative leadership is not a one-style-fits-all solution and is never the fastest way to proceed.

Figure 11.3 Levels of Participation and Collaboration

| COERCION: PRESENT BUT NO INPUT | PARTICIPATION: JUST ENJOYING THE RIDE | COOPERATION: I'LL WORK ON YOUR GOAL | COLLABORATION: ALL COMMITTED TO *OUR* GOAL |

It takes a special kind of leader, with a developed emotional intelligence, to be able to effectively collaborate with people who represent a broad diversity of expertise, experiences, cultures, ages and geographies. Often cited good team management practices that are likely to optimise effective collaboration, are:

- **Purposed collaboration:** -Making sure that the aims and objectives for the collaboration are clear.

- **Facilitating broad engagement:** -Ensuring the collaborative project team leader recognises it is not their role to come up with the 'right' idea; rather, they are there to establish and maintain a collaborative process, that allows everyone to participate fully in the group's work, ensuring all participants get heard, soliciting the opinions of those who haven't spoken.

- **Establishing group norms** (e.g. respect, participation, trust, confidentiality, handling of IP, conflict resolution, etc.).

- **Always recognising peers' contributions** and efforts by having the confidence to share credit generously and not take it all for themselves.

To help develop managers' ability to operate in such ways, wise innovators consider:

- **Using assessment tools** so that managers have a self-awareness of their most natural leadership styles and which ones to develop.

- **Coaching managers** to practice operating in a collaborative way (in order to develop further their leadership style armoury). For example, by coaching on a diverse internal (cross-functional) or external task where the coached manager is required to demonstrate a revised way of working on an Innovation-focussed project.

- **Ensuring that leadership development** embraces training in the skills of good networking, understanding cultural differences, listening and influencing.

- **Performance assessment** that evaluates a manager not just on what they personally have achieved, but also about how much they contributed to the success of others outside their team.

Element 7.3:
MANAGE INNOVATION ENABLING ACTIVITIES

Most managers are instinctively comfortable with success that can be quantified with metrics, so why some organisations treat Innovation in a more ethereal way is unclear.

Leaders in wise innovators embed a set of measures that track not just the key business results being achieved, but also track the most important innovation enabling activities that they have put in place, such as those in Elements 1-7 of this book's Innovation Capability Assessment Framework. They will create a company-level, enabler-focussed, dashboard of around 8–15 metrics, (if any more granularity is needed, it is typically done at the functional level). They will then systematically review these data sets in order to understand and act on, the causes of trends and, when possible, compare their performance to that of competitors and/or 'best in class' organisations to identify capability gaps that should be closed.

Accounting of: Overall Innovation Achievements

To track the key business results being achieved, wise innovators will consider measures such as:

- Percentage of overall revenues, or margin obtained from new product/service introductions.

- Estimated risk adjusted value of the current Innovation project portfolio (value of what is in Ideate + Validate + Scale pipeline in Elements 3–5), including assessment of both the quality, and value, of any specific IP created, or likely to be created.

- Indicators of the achievement of their specific Innovation strategy. For example the outcomes of the, for them, most strategically key data sets from Element 1 below.

Typical Measures for Element 1: Innovation Strategy

To help assess the deployment and relevance of their Innovation strategy, wise innovators will use measures such as:

- Spend on Innovation as percentage of revenue or margin.
- Number of new markets entered and explored.
- Percentage increase in specifically targeted market shares, as a result of Innovation.
- Indicators of product or platform diffusion/adoption.
- Indicators of Innovation capabilities built.
- Achievement of targeted brand/image transitions.
- Employee and partner feedback on clarity of Innovation goals.
- Indicators of capital and other resources available to support ideas.

Typical Measures for Element 2: Data And Analytics

To track efficiency and effectiveness of their data and analytics, wise innovators will use measures such as:

- Indicators of data and knowledge breadth.
- Indicators of data and knowledge accessibility and sharing.
- # External knowledge sources/communities engaged with.
- Return on Investment (ROI) of key knowledge partnerships/collaborations.
- Training and expertise in analytical techniques.

- Percentage of order book assigned to learning customers (customers who will give insights and feedback).
- Number of opportunity briefs created.
- Clarity of opportunity briefs created.

Measures for Elements 3–5: Ideate/Validate/Scale

To track their investment in the Ideate + Validate + Scale pipeline, wise innovators will use measures such as:

- Number of opportunity briefs processed and met/achieved.
- # of ideas created (internal and external, from within units and cross units).
- # of ideas validated.
- # of ideas scaled to market successfully.
- Velocity (e.g. average time from idea to validate to market scaling) and costs of projects.
- Portfolio position vs. portfolio plan: is it balanced in high and low risk opportunities?

Measures for Element 6: Innovation Ready Culture

To assess their progress with developing and maintaining an Innovation Ready Culture, wise innovators will use measures such as:

- Culture/Climate indexes.
- # Collaborations in cross-functional teams and external teams.
- Percentage of employees for whom Innovation is a key performance goal.
- Indicators of employee participation in Innovation-related forums and systems.

- Indicators of organisation-wide Innovation capability.
- Indicators of Innovation-related training and development.
- # Rewards and recognitions.

Measures for Element 7: Future Focussed Leadership

To oversee development of Future Focussed Leaders, wise innovators will use measures such as:

- Percentage of leaders' overall time dedicated to Innovation/future business (i.e. a measure of ambidexterity).
- Percentage of executives' time coaching others.
- Percentage of leader time participating in specific Innovation or collaboration projects.
- Number of leaders in the company classed as 'entrepreneurs'.
- The 'value' of leaders personal networks.
- Number of managers with non-executive roles in other organisations.
- Effectiveness of leadership communications.

ASSESS YOUR CAPABILITIES

It may now be timely to assess your own organisation's performance with respect to this element. A good way of doing this is to form an assessment team and provide them access to the following 10 questions, which are designed to help you debate and assess your current capabilities and achievements.

How to Assess:

1. Review whether a data-gathering phase will be beneficial, prior to answering the questions, and if so whom to involve.

2. Have each assessment team participant, individually, complete an analysis of each question, circling a score of 1 if they believe that few relevant outcomes/actions can be demonstrated, scoring 5 if they believe you truly have role model practices/outcomes and interpolating between scores 2 to 4 for other conclusions.

3. Undertake a workshop with the aim to create a consensual analysis and score for each question and debate what you think is the next big thing for you to work on.

What to do with the Assessment:

1. Take an average of your agreed scoring of the ten questions and conclude your current 'Reboot Status' for this element. Use this status number to form a base-line for future assessments and to share with interested stakeholders.

2. In areas that you choose to act on, seek insights and good practices that may help you conclude what to do next. We use funds from the sale of this book to fund the Innovation Reboot Project at www.innovationreboot.org. After considering what is the next big thing for you to address, take a look at this website for knowledge, ideas and discussions on each capability element.

Q7.1:
Our **leaders are ambidextrous** and balance a valid focus on efficiency and cost, of the operations of today, with innovating for the future.

1. Unable to Demonstrate
2. In Parts
3. Basics Covered
4. Fully Able to Demonstrate
5. We're a Role Model

Q7.2:
At all levels, our leaders can clearly articulate **where we need Innovation** to drive growth and **how we plan to obtain it.**

1. Unable to Demonstrate
2. In Parts
3. Basics Covered
4. Fully Able to Demonstrate
5. We're a Role Model

Q7.3:
Our leaders understand when it is likely to be of value to **lead with a collaborative style.**

1. Unable to Demonstrate
2. In Parts
3. Basics Covered
4. Fully Able to Demonstrate
5. We're a Role Model

Q7.4:
Our leaders can effectively **facilitate, collaborate and coach.**

1. Unable to Demonstrate
2. In Parts
3. Basics Covered
4. Fully Able to Demonstrate
5. We're a Role Model

Q7.5:
Our **leaders are widely networked** (e.g. with competitors, consumers, Government officials, universities, conferences, non-executive directorships, etc.).

1. Unable to Demonstrate
2. In Parts
3. Basics Covered
4. Fully Able to Demonstrate
5. We're a Role Model

Chapter 11: Element 7 FUTURE FOCUSSED LEADERS

Supporting Evidence:

Supporting Evidence:

Supporting Evidence:

Supporting Evidence:

Supporting Evidence:

Q7.6:
Our leaders understand and **can use the Innovation tools and techniques** that we have in place.

1. Unable to Demonstrate
2. In Parts
3. Basics Covered
4. Fully Able to Demonstrate
5. We're a Role Model

Q7.7:
Our leaders **understand, value and optimise diversity.**

1. Unable to Demonstrate
2. In Parts
3. Basics Covered
4. Fully Able to Demonstrate
5. We're a Role Model

Q7.8:
Our leaders **use feedback from employees and partners** in order to review our leadership behaviours and actions.

1. Unable to Demonstrate
2. In Parts
3. Basics Covered
4. Fully Able to Demonstrate
5. We're a Role Model

Q7.9:
We use a **balanced set of measures to manage Innovation enabling practices** and oversee our desired **outcomes/results.**

1. Unable to Demonstrate
2. In Parts
3. Basics Covered
4. Fully Able to Demonstrate
5. We're a Role Model

Q7.10:
Compensation **of key leader(s) is related to accountability** for our strategic Innovation aims.

1. Unable to Demonstrate
2. In Parts
3. Basics Covered
4. Fully Able to Demonstrate
5. We're a Role Model

Supporting Evidence:

Supporting Evidence:

Supporting Evidence:

Supporting Evidence:

Supporting Evidence:

SECTION 3
APPENDICES

APPENDIX 1

WE ARE WHAT WE SHARE

To help us develop the Innovation Reboot Project and to allow you to build-on and use the insights in this book, you can go to the Innovation Reboot Project web site and download copies of the 7-element Innovation Capability Framework diagram. Over time additional materials will be made available.

The 7–element Innovation Capability Framework diagram (the diagram only) is openly shared under Creative Commons Attribution CC BY-SA Attribution-ShareAlike terms.

Innovation Reboot Framework

1. INNOVATION STRATEGY
2. DATA AND ANALYTICS
3. IDEATE
4. VALIDATE
5. SCALE
6. INNOVATION READY CULTURE
7. FUTURE FOCUSSED LEADERS

CURRENT STATE → DESIRED STATE

Licensed by www.innovationreboot.org

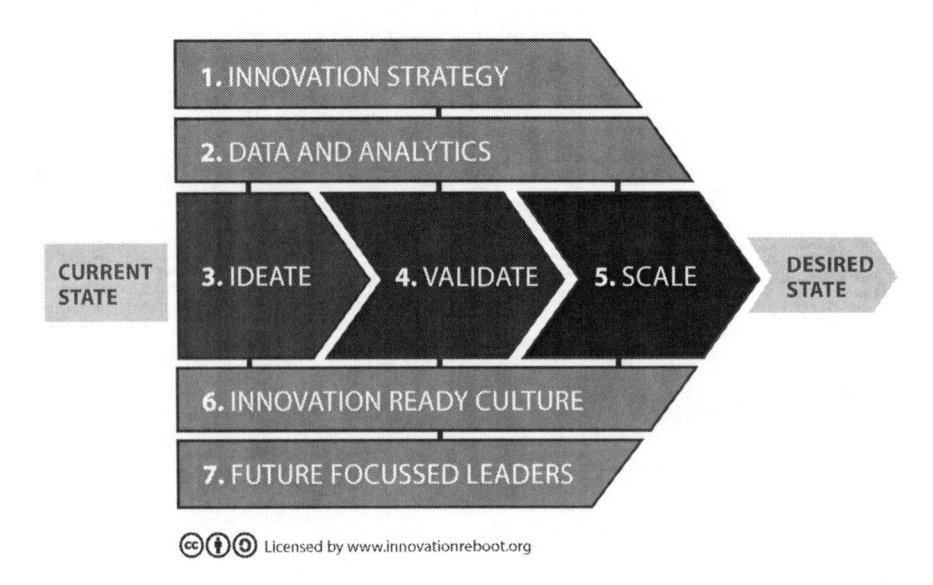

This means that the diagram, in its format above, is immediately available for royalty free use provided the shown acknowledgements (the three Creative Commons symbols and the reference to 'Licensed by www.innovationreboot.org') are maintained in <u>all</u> places of reprint, amendment or use.

Full licence terms can be found at:
http://creativecommons.org/licenses/by-sa/3.0/

APPENDIX 2

WHAT NEXT

Leadership Agenda (the Publisher of this book) are a founding sponsor of the Innovation Reboot Project and encourage you to get involved with it.

Over time it is foreseen that the Innovation Reboot Project will evolve to include:

- **Ongoing sponsorship opportunities** to promote Innovation good practice sharing and/or fund Innovation related research.
- **Rights to reprint and use Innovation tools**, for example, for a small donation on the web site you can obtain a Creative Commons CC BY-SA royalty free licences for an electronic, printable, file copy of the questions in this book.
- **Opportunities to collaborate.**

Articles, that share good practice or provoke constructive debate, can be posted on the website to one, or more, of the seven Innovation Capability Framework elements. Full author acknowledgements are given.

There are simple and non onerous author guidelines which are that any article submitted should be:

- In English.
- Between 200 and 1,000 words (pictures or diagram maybe be also submitted).
- Not overtly advertorial.
- A substantiated, honest ('warts-and-all') view or insight that is capable of adding value to the building body of knowledge.
- Able to be openly shared, under a Creative Commons, Attribution CC BY-SA Attribution-ShareAlike licence.

Go to **www.innovationreboot.org.** to see updates on progress with this project and opportunities to get involved.

APPENDIX 3

NOTES

Chapter 1: **DEFINITION AND SCOPE OF INNOVATION**

[*1] Lafley, A. G., and Martin, R. L., '*Playing to Win: How Strategy Really Works*'. First Edition. Harvard Business Review Press, 2013 and in associated HBR blogs.

Chapter 2: **COMMON IMPERATIVES**

[*1] Looking to 2060: Long-term growth prospects for the world. '*Economic Policy Paper No. 3*', November. 2012, OECD, Paris.

[*2] Zikopoulos P. C. and Eaton C., 2012, '*Understanding Big Data*', IBM, McGrawHill.

[*3] United Nations Millennium report, '*The role of the United Nations in the 21st century*', 2000, United Nations, New York.

[*4] Manyika J., Chui M., Bughin J., Dobbs R., Bisson P., and Marrs A., *'Disruptive Technologies: Advances that will transform life, business and the global economy'* McKinsey Global Institute, McKinsey Quarterly, May 2013, Stamford.

Chapter 5: ELEMENT 1 INNOVATION STRATEGY

[*1] Ansoff, H. I., *'Strategies for Diversification'*, Harvard Business Review, Vol. 35 Issue 5, Sep-Oct 1957, pp. 113-124.

[*2] Christensen, C. M., 1997 and 2011. *'The Innovators Dilemma'*, Boston, Harvard Business School Press and Harper Business reprint and Christensen, C. and Raynor, M., 2003. *'The Innovators Solution'*, Boston, Harvard Business School Press.

[*3] Nagji, B. and Tuff,G., May 2012, *'Managing Your Innovation Portfolio'*, Boston, Harvard Business Review. Reprint R1205C. with origins in Baghai, Mehrdad, Coley and White. *'The Alchemy of Growth: Practical Insights for Building the Enduring Enterprise'*. New ed. Basic Books, 2000.

[*4] Kim, C. K. and Mauborgne, R., 2005. *'Blue Ocean Strategy: How to Create Uncontested Market Space and Make the Competition Irrelevant'*. Boston, Harvard Business School Press.

Chapter 6: ELEMENT 2 DATA AND ANALYTICS

[*1] Drucker, P. F. *'Innovation and Entrepreneurship'*. 2Rev Ed. A Butterworth-Heinemann Title, 2007.

Chapter 7: ELEMENT 3 IDEATE

[*1] Stevens, G. A. and Burley, J., 1997. *'3000 Raw Ideas = 1 Commercial Success!'* Research-Technology Management 40(3), pp.16-27.

Chapter 8: **ELEMENT 4 VALIDATE**

[1] Merrill, P. *'Innovation Generation: Creating an Innovation Process and an Innovative Culture'*. American Society for Quality, 2008.

[2] Rogers, E. (1995). *'Diffusion of Innovations. (4th edn)'*. New York, NY: The Free Press.

Chapter 11: **ELEMENT 7 FUTURE FOCUSSED LEADERS**

[1] Dyer, J., Gregersen, H., and Christensen, C., *'The Innovator's DNA: Mastering the Five Skills of Disruptive Innovators'*. Brilliance Corporation, 2012.

APPENDIX 4

GLOSSARY of TERMS

The meaning of key Innovation-related terms, in the context that they have been used in this book:

Agility:
An ability to rapidly and efficiently adapt and respond to changes.

Business Model:
The components of the business that, together, create and deliver value to stakeholders.

Capabilities:
An ability to achieve a specific goal or task that is brought about by the timely provision of both competence (knowledge, expertise, etc.) and capacity (available resources to complete the task).

Capacity:
A measurement of the resources available to complete a task – in terms of quantity, size, volume or other number.

Climate:
The way things are seen and done in an organisation and the pervasive behaviours and attitudes that characterise it.

Cloud Computing:
A way to access scalable computing resources through the Internet, often at lower costs as the resources are shared across many users.

Co-creation:
The practice of developing offerings through the collaborative efforts of multiple stakeholders, typically involving stakeholders from outside the organisation (for example, customer co-creation).

Competence:
A valid mix of knowledge, skill and experience, leading to the ability to effectively complete a task. The term can be applied to individual competence(s) or collectively looked at as an organisation-wide cluster of abilities, that come together to form a pervasive area of specialised expertise, that may be a root of competitive advantage for an organisation.

Concept:
A promising product/service/business idea prior to its introduction to the market.

Corporate Governance:
A framework of authority and control within an organisation used to help it fulfil its legal, financial and ethical obligations.

Creativity:
The generation of ideas for new or improved products, services,

processes, systems or social interactions. It can be said that creativity is thinking up/about new things; Innovation is doing new things.

Crowd funding:
Financial resources obtained as small contributions from a large number of sources.

Culture:
The values, norms and beliefs that are shared by people and groups in an organisation.

Data:
Observations or facts in raw or unorganised form.

Discounted Cash Flow (DCF) Analysis:
A method of valuing a project, company, or asset using the concept of the time value of money. DCF is used to calculate the value of future cash flows in terms of an equivalent value today. All future cash flows are estimated and discounted to give their net present values (see NPV).

Early Adopters:
A segment of customers who are inclined to buy into new product/service offerings at an early stage in their life cycle.

Empowerment:
The process and rules by which individuals or teams are able to take decisions and operate with a degree of autonomy.

Foresight:
The ability to detect/predict significant upcoming change in a way that enables the prediction to be used in long-term decision-making.

Good Practice:
Methods, tools or techniques that have been shown to deliver superior performance.

Hypothesis:
A provisional conjecture to guide investigation.

Idea:
An original thought, or a thought not previously associated to the problem being considered; something to try out, explore and experiment.

Information:
Raw data that has been analysed and organised for a purpose and presented in a way that gives it meaning and relevance.

Innovation:
The practical translation of ideas into new products, services, processes, systems, business models or social interactions.

Innovation Strategy:
A high-level plan that describes short- and long-term aims and tactics for growth through the development of new, or significantly changed products, services, processes, systems, business models or social interactions.

Insight:
An understanding of the forces behind an action or its outcome.

Knowledge:
An understanding or belief about a subject, which is justified by supportive alignment of accumulated data, information and experiences.

Mission:
A statement of the purpose or 'raison d'être' of an organisation.

Moore's Law:
A widely used truism attributed to the Intel co-founder Gordon Moore. He observed that it has been a consistent fact that the number of transistors that can be placed on a computer chip (and thereby its computing

power), doubles every two years. This means that in most technology sectors, the available processing power doubles (or costs halve) every two years.

NPV:
The Net Present Value of the future cash flows predicted for an investment, less the cost of that investment (often used within a Discounted Cash Flow approach).

Organisational Agility:
The ability of an organisation to respond and adapt, in a timely way, to an emerging threat or opportunity.

Partnership:
A durable working relationship between an organisation and external parties that is created with mutual objectives and sustained benefits in mind.

Perception:
The opinion, be it intuitive or factually based, that an individual or, in business, that a stakeholder has formed.

Stakeholder:
A person, group or organisation that has a direct or indirect interest in the organisation and can either affect the organisation or be affected by it.

Strategy:
A high-level plan describing the tactics by which an organisation intends to achieve its mission and vision.

Trend:
A change that will have a significant enduring macro impact.

Values:
Philosophies or principles that you believe should guide conduct.

Value Proposition:
A promise of value to be delivered and a belief from a customer (or other stakeholder) that value will be experienced.

Vision:
A description of what an organisation is attempting to achieve in the long-term.

APPENDIX 5

ACKNOWLEDGEMENTS

I would like to acknowledge and thank all those who have assisted, persisted, influenced, or generally helped to maintain my sanity during the production of this book. Particular thanks are due to all friends and colleagues at EFQM and other global or regional business excellence awards administrations, with whom the learning journey has always been a two-way experience and, for me at least, a great pleasure. This book is a collection of proven practices, based, in many cases, on these privileged experiences.

Finally, thanks are due to Jo, Jane and Tom, my immediate family, for allowing me the (mainly uninterrupted) gift of time to pursue my passions and launch this learning experiment.

APPENDIX 6

ABOUT THE AUTHOR

Chris Hakes started his working life in the research labs of mega corporations, then became a Chartered Engineer and rose through the ranks to hold profit-accountable regional management roles. He subsequently escaped and has become an author, management heretic, educator and veteran/founder of several start-up businesses, including university spin-off's, social enterprises and projects in the public sector.

Through a long association with EFQM and other operators of global, national and regional business excellence award schemes, Chris has had the privilege of working with and assessing the performance of some of the world's leading organisations, in both public and private sectors. After supporting the launch of EFQM in 1988, Chris worked with them for over two decades as a part-time lecturer in the so-called EFQM Faculty. He was part of the team that developed the first version of the EFQM performance model over 20 years ago and has assisted EFQM model review teams ever since. He has authored several internationally distributed publications on Practical Excellence, Future Focussed Leadership and Organisation Self-assessment, in addition to producing tailored learning materials, in both paper and electronic form, for several international corporations.

In his current working life he spends his days split between coaching start-up organisations, authoring books and management education.

Chris lives near Cambridge in the UK and can be reached on LinkedIn at **http://uk.linkedin.com/in/chrishakes/**.

INDEX

A

adjacencies
    for growth, 51-56
    competitor, 69
agility
    in processes, 118-119
    hyper-agility, 125-126
    definition, 171
Amazon, 24
ambidexterity, 145-146
ambitions
    general, 25-34
    strategic, 47-56
analysis
    strategic tools, 48-49
    data basis, 65-72
    bias, 72
Apple, 55, 122, 124

assessments
    innovation capability, 36-38
    innovation strategy, 59-63
    data and analytics, 77-81
    ideation, 95-99
    validation, 111-115
    scaling, 127-131
    innovation ready culture, 139-143
    future focussed leadership, 155-159
assumptions
    hypotheses, 46, 57
    product and service, 93
    people, 94

B

behaviours
    leadership, 37, 145-153
    purchaser, 109
    effect on culture and climate, 135-137, 172
beliefs
    link to decision bias, 72
    effect on culture, 133-135, 173
bias, 67, 72-73
breakthrough, 56
budget(s), 24, 46-47, 55-56, 71, 74, 94, 103-110
business plans, 105-110

C

capability
    Innovation Capability Framework, 14
    capability gaps, 31-34
    capability and governance, 35-39
    building capability, 46
    definition, 171
Christensen, C.M., 54, 148
cloud
    data, 27
    computing, 71

climate (definition), 172
coaching, 120, 138, 148, 149, 153
co-creation, 75, 172
collaboration
    in external ideation, 87
    collaborative leadership, 145-149
commoditisation, 27, 54
communities, 85, 151
competence
    gaps, 33
    core competence, 69, 105
    for acquisitions, 88
    of people, 134-138
    definition, 172
competition
    nature of, 27-28
    data on, 67-72, 150-153
    when validating, 102-105
    when scaling, 125-126
concept
    validation, 101-110
    definition of, 172
consensus
    decision making skills, 72-75
    in workshops and governance, 37-37
Cook, Tim., 122
core
    business, 47-48
    idea, 74, 84
    competence, 69, 105
CreateSpace, 24
creativity, 83-89, 105, 172-3
crowd funding (definition), 173
culture
    organisational, 133-138,
    effect on ideation, 83-89
    impact when scaling, 121-124
    role-modelling, 146-149
    definition, 173

customer(s)
    data for strategy, 48-57, 67-69
    engaging in ideation, 87
    feedback when validating, 103-110
    feedback when scaling, 125-126

D

dashboard, 120, 125, 150
data
    ubiquitous data, 27-30
    horizons, 67- 68
    analysis and forecasting, 47, 69-72, 106
    sharing, 73 -76
    when scaling, 120-125
    management data, 150-153
    definition, 173
decisions
    factual basis, 67-71
    bias, 71-72
    idea selection, 89-94
    in business plans, 105-110
    collaboration in, 148-150
definition (of innovation), 19- 24
diffusion, 28, 108-110, 151
disruption
    in markets, 29-30
    disruptive innovation, 48-56
    management of, 148
diversity
    effects in ideation, 83-89
    when recruiting, 121-124
    leadership with, 147-150
Drucker, Peter, 71-72

E

early adopter, 109-110, 173
economies, 26-27
ecosystem, 102

EFQM, 12, 21, 177, 179
enablers
    innovation enabling activities, 150-153
    enablers of culture, 138
entrepreneur, 12, 71, 106, 117, 121, 123, 153

F

Facebook, 29, 89, 124
feedback, 137-138
forecast(ing), 27, 32-33, 46, 66, 69-72, 93, 108, 109-110
foresights, 23, 69-75, 173

G

GDP, 26, 27
GlaxoSmithKline, 52, 55
globalisation, 26-30, 68
Google, 29, 126
Gore (Goretex), 138
governance, 35-39, 108, 121, 172
growth
    of organisations, 20, 24, 26, 33
    of data, 27, 67
    growth opportunity analysis, 48-57
    market diffusion and growth, 109-110,
    process alignment, 119-121
    people alignment, 121-124

H

Hakes, Chris., 15, 179
Hoffman, Reid, 124
hyper-agility, 125-126
hypotheses, 57, 74, 93, 104, 100, 174

I

ideas
    defining the opportunity, 74-75
    ideation, 83-87
    ideas management, 88-89

ideas selection, 90-94
idea validation, 101-110
idea scaling, 117-126
definition, 174
IdeaScale, 89
imperatives, 25-34, 47, 134-136
incubate, 37, 93, 101-102
indicators, 110, 125, 150-153
information, 70-71, 174
Innovation Capability Assessment Framework, 14
insights
    related hypotheses, 57
    use of data, 67-76
    customer, 125, 152
    definition, 174
integration, 45-47, 67, 84
investment, 48-57, 65, 85, 151-152, 175
iTunes, 55-56

J

Jobs, Steve., 124

K

Kaggle, 87
Kim, C.K., 56
Kindle, 24
Knowledge, 26, 33, 38, 70-73, 89, 124, 151, 174

L

Lafley, A.G., 20
leadership
    ambidexterity, 147-148
    collaboration style, 148-149
    innovation governance, 38-39
    management of enabling activities, 150-153
LinkedIn, 124
Lucozade, 52, 55

M

margin, 150-151
market maturity model, 108-11
matrix
    opportunity analysis, 48-53
    idea evaluation, 92-94
Mckinsey, 29-30
measures of innovation, 150-153
minimum viable product, 104
mission, 174-175
modelling, 71, 137
morph, 126

N

networking, 88, 126, 148, 149
Nike, 56
Nintendo, 55

O

OpenIDEO, 87
opportunities
    for innovation and growth, 22, 26-27, 29
    in data, 66-75
    for managers, 121-124
    to participate, 135-138
    for leaders, 147-150
    opportunity analysis matrix, 48-56
    opportunity brief, 73-74, 90-93, 106, 137, 152
outcomes (of innovation), 21-24

P

partnership
    as stakeholder, 69
    as collaborator, 87-88, 148
    definition, 175
people
    and ideation, 93

in business plans, 107
   when scaling, 119-126
   culture, 134-138
pipeline, 33, 39, 46, 47, 73, 93, 105, 150, 152
pivot(ing), 119, 126
planning
   forecasting, 39, 45-57, 66, 102-110
   data in planning, 66-75
   validation plans, 101-111
portfolio, 31, 33, 36, 46-48, 105, 150, 152
process(es)
   as a category of innovation, 23-23
   governance, 38-39
   strategic, 47-48
   for idea vetting, 92
   tight-loose management, 105, 119-121
   for collaboration, 148-150
Proctor and Gamble, 20
proposition, 23, 93, 176

R

recruitment, 121-124
resources, 26, 33, 47, 54, 67, 85, 88, 101, 105, 108, 151
revenue
   gaps, 31-34
   growth, 47, 50, 55, 108-110
risks, 26-30, 31, 51-56, 72, 104, 108, 136
Rogers, E., 108-109

S

scale
   global, 25-29
   scaling of ideas, 117-126
   measures of, 152
Schumpeter, Joseph, 21
scope
   of innovation, 19-24
   of data and analytics, 67-71

scorecards, 105
social
    trends, 66-72
    networking technologies/media, 67-68, 87-88, 125-6, 126
    legislation, 108
Spigit, 89
stakeholder definition, 175
strategy
    for innovation, 45-57
    measures, 151
    definition, 175
'sustaining' (category of innovation), 30, 50-54

T

technology
    as an imperative for change, 26-30
    considerations for strategy, 45-57
    data on, 69, 74
    use to enable creativity, 85
transformation, 24, 30, 50-56
trends
    observation and management, 69-72
    trend-owners, 72
    intersections, 72
    definition, 175
Twitter, 29, 89, 126

U

Universities
    academia and, 12
    research, 69

V

validate, 101-110
values
    organisational, 120-124, 138
    definition, 176

vision
 use of, 47, 120, 136, 175
 definition, 176

Y

YouTube, 126

Z

Zuckerberg, Mark, 124

Lightning Source UK Ltd.
Milton Keynes UK
UKOW04n2259280713

214483UK00002B/2/P